Also by Jay Ingram

The Science of Why
Volume 5

Answers to Questions About the Ordinary, the Odd, and the Outlandish

Jay Ingram

PUBLISHED BY SIMON & SCHUSTER

NEW YORK   LONDON   TORONTO   SYDNEY   NEW DELHI

Simon & Schuster Canada
A Division of Simon & Schuster, Inc.
166 King Street East, Suite 300
Toronto, Ontario M5A 1J3

This Simon & Schuster Canada edition November 2020

SIMON & SCHUSTER CANADA and colophon are trademarks
of Simon & Schuster, Inc.

For information about special discounts for bulk purchases,
please contact Simon & Schuster Special Sales at 1-800-268-3216
or CustomerService@simonandschuster.ca.

Illustrations by Tony Hanyk (tonyhanyk.com)

Manufactured in the United States of America

10   9   8   7   6   5   4   3   2   1

Library and Archives Canada Cataloguing in Publication

Title: The science of why. Volume 5 : answers to questions about the ordinary,
    the odd, and the outlandish / Jay Ingram.
Names: Ingram, Jay, author.
Description: Simon & Schuster Canada edition.
Identifiers: Canadiana (print) 20200202308 | Canadiana (ebook) 20200202316 |
    ISBN 9781982140854 (hardcover) | ISBN 9781982140861 (ebook)
Subjects: LCSH: Science—Popular works. | LCSH: Science—Miscellanea.
Classification: LCC Q162 .I555 2020 | DDC 500—dc23

ISBN 978-1-9821-4085-4
ISBN 978-1-9821-4086-1 (ebook)

To the memory of Penny Park, a longtime friend and colleague, who knew more about science journalism than just about anybody and brought to it a determined ethical stance

Contents

Part 3: Oddities and Eccentricities

Part 4: Perplexing Phenomena

The Science of Why
Volume 5

Part 1
Awesome Animals

Are octopuses from outer space?

In 2018, THIRTY-THREE SCIENTISTS published an article in the journal *Progress in Biophysics and Molecular Biology* arguing that octopuses are actually alien life-forms that arrived on earth on a space rock 270 million years ago.

Yes, you read that correctly.

And, yes, it's as crazy as it sounds. Octopuses are unique in so many ways that one scientist has said, "It's like meeting an intelligent alien"—but he didn't mean that literally.

This theory is an attempt to explain why the octopus is so different, especially genetically, from even close relatives like the chambered nautilus. But it's not just about the genes: octopuses have a bit of an alien look. They're also extremely smart, can change pattern and color to blend into their surroundings, and each of their eight arms has its own brain. That last fact alone puts the octopus in a class by itself.

Octopuses have a legendary ability to escape aquariums. They can use tools and solve problems, and they seem able to think in an almost humanlike way. One well-known experiment tested octopuses on their ability to open a complicated set of boxes to get at a crab treat inside. The first box had a latch that needed to be twisted open; inside that was another box, which slid to open; and that in turn was inside a third box, which had two different locks. Two or three trials was all octopuses needed to be able to open all the boxes in three or four minutes.

What makes an octopus so smart? Its nervous system has 500 million neurons, ranking it somewhere between the European rabbit and the western tree hyrax, an African guinea pig–like animal. But the number of neurons alone isn't a good measure of intelligence. The way those neurons are organized is important, too, especially in the octopus.

Of the octopus's 500 million neurons, 150 million are found in the brain, and the other 350 million are shared among the arms. In effect, each arm has its own minibrain and is capable of making its own decisions. If one of the suckers detects something delicious, for example, that arm will alert the other arms to what's happening. Then the arm will curl around the food, making a kind of hand, while the rest shapes itself into an upper and lower arm and an elbow so that the "hand" can bring the food to the mouth.

Science _Fact!_ _Eerily, an arm that's been separated from an octopus's body will still grab food and try to pass it to where the mouth should have been. It will also try to crawl away on its own._

What sets octopuses apart? The chambered nautilus, a cousin, is not nearly as intelligent. Some experts think the key difference is that the nautilus never lost its shell. It leads a relatively safe and lengthy life (it can live up to twenty years), but perhaps not the most exciting one. The octopus, on the other hand, lost that shell completely as it evolved into its modern form. The argument is that losing the protection of the shell put evolutionary pressure on the octopus to become smart, agile, well camouflaged, and capable of squeezing into the tiniest spaces. The trade-off is that it typically survives only two or three years in the wild.

I'm a sucker
for brains.

All of these features make the octopus an extraordinary creature, but the scientists arguing for its alien origin concentrate only on its genes. Genes build animals, but octopuses do it in a different way. They are genetically nimble and can apparently alter the way their genes are expressed, allowing them to respond rapidly to changes in their environment. But that nimbleness reduces the octopus's ability to fine-tune its DNA over longer periods of time, the process that drives evolution. This machinery is not unique to octopuses—we humans have it, too—but in most other species, it's a tiny feature of the genome, whereas in the octopus it is absolutely crucial.

That brings us back to the "octopus as alien" theory. The argument is that this ability to revamp the genome is not typical of other animals on Earth—therefore octopuses must have come from space. They would have arrived not as full-grown animals but as frozen octopus eggs. It would be pretty cool if it were true, but the evidence isn't exactly airtight. It may look like the octopus appeared out of nowhere 270 million years ago, but there are so few octopus remnants in the fossil record that it's impossible to be sure. And while their practice of gene editing is intriguing and unusual, does the fact that we do it, too, mean we come from space as well?

If you still want to believe that octopuses are aliens (and I wouldn't blame you if you do!), you should know that for decades now the authors of this paper have been pushing the idea that life arrived on earth from space. *All* life, that is. So far they haven't been able to persuade the rest of the scientific community of that.

Why are porcupine quills so hard to pull out?

You MIGHT HAVE SEEN one of those online images of an unfortunate dog with his muzzle full of porcupine quills. It wasn't a great experience for the dog when those quills went in, but it was certainly much worse when they came out! From the porcupine's point of view, though, the quills are one of the most advanced defense systems in the natural world.

Porcupines are the world's third-largest rodent, after the beaver and the capybara. There are species scattered around the world, but the several subspecies in North America are unique for their barbed quills. Each quill is about 11 centimeters (4.3 inches) long, and the top 40 percent or so, the black part, is covered in barbs. One porcupine can have as many as thirty thousand quills on its body and tail!

I like to needle my enemies.

If you're thinking the barbs must make the quills really hard to pull out, you're right: as you tug on a quill to remove it from your dog (or yourself!), the barbs, which are usually flat against the shaft, open up and spread out horizontally into the flesh, a little like an umbrella being opened. The harder you pull, the more they spread out, until eventually the quill is thicker and wider than the puncture hole it made in the first place.

But the barbs don't just make the quills hard to remove—they also make them more able to *penetrate* the flesh of an animal. Quills have been found deep within other species' muscles and in just about every organ: stomach wall, liver, lungs, and kidneys. It isn't clear why the barbs aid penetration, but the experimental data are unambiguous. A team led by biomedical engineer Jeffrey Karp at MIT compared barbed porcupine quills with hypodermic needles, quills with the barbs sanded off, barbless quills from the African porcupine, and even artificial quills fashioned from polyurethane. Careful measurements of the penetration force needed for each of these showed conclusively that barbed quills pass through the flesh most easily—even better than a standard 18-gauge hypodermic needle. What's more, the barbed quills cause less tissue damage going in.

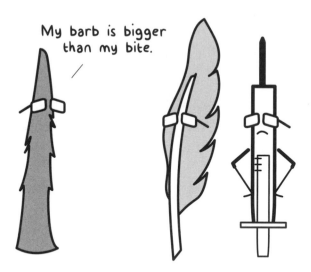

Karp and his colleagues concluded that the relative lack of damage was related to how the stress of quill entering flesh is distributed. They argued that because that stress is concentrated around the barbs, a quill operates more like a serrated knife than a nail, entering with less force and making a cleaner cut. You can see it in photomicrographs: barbed quills cut much more smoothly than the barbless versions. Even artificial barbed quills made of polyurethane needed 35 percent less energy to penetrate muscle than artificial barbless quills. This doesn't mean they don't hurt, though. They really do.

Did You Know . . . Once a porcupine has embedded dozens of quills into the body of an attacker, how does it get away? After all, the quills are attached to the porcupine, too. It's a serious issue because if enough quills are stuck in an enemy, the force required to pull away may be too much for the porcupine. But over time, a solution has evolved. When the porcupine smacks its target, the impact momentarily drives the quills back, breaking the links in which they're seated. Then separation is easy. What's extra cool about this is that the impact has to be powerful, like a porcupine lashing out with its tail. If the animal just lies down on a tree limb, that won't be enough to dislodge the quills.

Dr. Karp and his teammates at MIT have biomimicry on their minds: they're working on an improved hypodermic needle that would penetrate as easily as the barbed quill. They've already tested a prototype of a polyurethane needle with barbs, and it requires 80 percent less force than a barbless needle. Of course, a needle that goes in as easily as a porcupine quill can't also be as hard to remove! Karp and his colleagues think the answer to that problem may lie in creating barbs that change shape once wet. But resistance to being pulled out could be an advantage, too: they're also thinking beyond needles, to wound protection, to a patch to remain in place over tissues as they heal. In this case, both attributes of barbed quills would be important, especially the resistance to detachment, as long as it can be controlled.

In the meantime, there are plenty of biological questions to consider, like why does the North American porcupine have barbed quills while its African cousin doesn't? It's certainly not for lack of danger. The African porcupine's main predators are lions, hyenas, and large birds of prey. So is there something different or more intense about the predatory pressure on our local porkies that pushed them to evolve barbed quills?

In some ways, the quills are more of a warning than a first line of defense, and the porcupine diligently advertises them by raising them in times of danger, creating an almost skunk-like white stripe down its back. It will also clatter its teeth together and release a pungent odor to announce it's ready to use its quills. The odor has been described as "the smell of goat or perhaps an exotic cheese." The one thing a porcupine doesn't do is throw its quills. Naturalists have known this for 150 years, but the rumor still persists—perhaps because the quills release so easily when striking a target that they give the impression they flew through the air.

Unfortunately, sometimes all the brilliant engineering of barbed quills isn't enough. Although porcupines can live into their twenties, they seldom get that chance. There are several porcupine predators in North America—including the fisher, cousin to the weasel, which approaches the animal from its defenseless front. Barbed quills are a sophisticated deterrent, but they're not quite perfect.

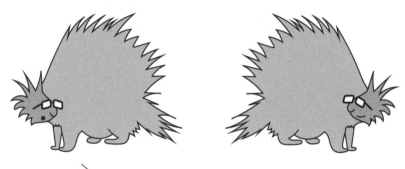

Have you felt the
kickback on this thing?

Can bees count?

Scientists have known for a long time that many animals can count, and some are remarkably good at it. Alex, an African gray parrot trained for years by Dr. Irene Pepperberg, understood numbers up to 6 no matter the color or shape of the objects he was counting. Alex, chimpanzees, and rhesus monkeys have all shown an ability to associate a numeral with a specific number of objects. When you realize that this means they can look at the symbol 5 and know it means five objects, it's startling. But the idea that bees can do the same? That's almost unbelievable.

Several experiments by Scarlett Howard, an expert in animal cognition, and a team of French and Australian researchers have shown that bees, too, can learn the rudiments of numbers. But Howard and her team started with something simpler: teaching

One, two, bee.

bees to grasp the difference between the numbers 4 and 5, 4 and 6, 4 and 7, and 4 and 8. (The larger the number being compared to 4, the easier it is for the bees. The slight difference between five objects and four makes it hard for bees to tell them apart.)

A maze in the shape of the letter Y was the classroom. Bees entered the maze through a small hole in the stem of the Y, and when they reached the two arms, they could choose the one that had a sign with four objects (mostly squares, circles, or triangles of different sizes) or the one with a sign of more than four. In one version of the experiment, bees that chose the arm with four objects received a reward, a drop of sugar water. Those that chose the wrong arm received no reward. In a second version, the bees were given a reward for the correct choice and a punishment, a drop of bitter tonic water, for the incorrect choice. Bees in the second trial learned much more effectively. (This should not be taken to mean that all learning should be accompanied by punishment—they're bees!)

Did You Know . . . Howard's approach of teaching bees by rewarding them with sugar water or punishing them with tonic has been used in other kinds of experiments, too. In Australia, for example, honeybees were taught to distinguish between paintings by the French impressionist Claude Monet and the Australian Aboriginal painter Nonggirrnga Marawili. Researchers placed a tiny vial of tonic water in the middle of four Monets and a vial of sugar water amid four Marawilis. Within a single afternoon, the bees had learned which paintings to approach to get a treat.

So bees are good! They might not be as skilled with numbers as brainy parrots or higher primates, but they have only a million neurons in their brains, whereas parrots and primates have thousands of times more. (We humans have about 86 *billion.*) Does that matter, though? It's tempting to compare figures and be even more wowed by the bees' accomplishment, but some scientists think that correlating smarts to the number of neurons may not make much sense.

Scientists at Queen Mary University of London have argued that to distinguish four objects from six or seven, bees may need only a tiny number of neurons. They've even claimed that a computer-based neural network with just four "neurons" can do what the bees do. This is because it's more of a visual and memory trick than actual counting. They argue that if a bee flies over the objects it's supposed to count, scanning them and remembering the number of edges and shadings, the bright and the dark, a simple ensemble of neurons could guess the total number. In this scenario, the bee isn't really counting one, two, three, but is instead recording and storing a host of visual information, at least for a few seconds.

Questions about the nature of the bees' ability haven't slowed the research, though. Scarlett Howard and her team extended the maze study in a dramatic way, presenting their test subjects with a new and more complex challenge. This time when they arrived at the Y-shaped maze, the bees were confronted by a sign bearing either an N or an upside-down T. The N represented 2, the ⊥ 3. The task was no longer to match objects to objects, but a symbol (which in itself can mean anything) to a number of shapes. When they'd absorbed the symbol, the bees moved forward to choose between the arm with two objects at the entrance and the arm with three. If they'd just seen an N, they were supposed to choose two; if they'd seen a ⊥, they were to choose three. With enough training, the bees learned the task—after fifty trials, they were accurate 80 to 90 percent of the time.

Two bee or not two bee.

But even though they'd learned that N equals 2 and ⊥ equals 3, they couldn't reverse the process. That is, if faced with shapes at the entrance to the maze and the two symbols inside, they couldn't figure out what to do. So they have some limits, but still, matching a symbol to a number is significant.

Honeybees actually have an enormous repertoire of impressive brainwork. If several landmarks (this is what researchers call the miniature tents they use in their work) are laid out in a field, and bees learn that there's always sugar water between landmarks 3 and 4, that's where they'll consistently go. How do they do that? In some way, they're counting the landmarks. If the researchers move the sugar water to between landmarks 2 and 3, some bees will still search between 3 and 4, but others will home in on the new location. Adaptable.

Imagine this: bees learn to associate an odor with a particular sugar feeder. That memory of the place is so vivid that if the same odor is pumped into the hive, bees will go to that location, even if there's no sugar there anymore. They remember it was a good place, but they soon realize it isn't anymore.

Of course, scientists have thought about pushing the limits with bees. Could the insects, for instance, discriminate among *more than* two symbols with associated numbers? The problem is time. Each new step requires more training and a more elaborate maze system, and bees are just not that long-lived—maybe a few months. Such experiments would likely run out of subjects before they could yield results.

So the answer to the question "Can bees count?" is obviously yes. But a better question would be "Can they do square roots?"

Why do cowbirds lay their eggs in other birds' nests?

BECAUSE IT SEEMS LIKE THE PERFECT STRATEGY! Look at the alternative: most birds migrate north in the spring; find a nesting site; gather materials like twigs, string, and stray bits of cloth; build a nest; lay eggs; incubate the eggs while guarding them against predators; feed the hatchlings—and then, if there's any time left in the summer, do it all over again.

What are cowbirds doing all that time? Searching for existing nests to lay their eggs in, then eating and relaxing. You'd think more North American birds would have opted for this strategy, called nest (or brood) parasitism, but it's not as easy as it sounds.

In North America, there are two species of cowbirds: the bronzed and the brown-headed. The brown-headed cowbird, by far the more common of the two, was already well established on the prairies centuries ago. Flocks of these birds followed the huge mobile herds of bison, which disturbed the grass as they moved, releasing hordes of insects and seeds that the cowbirds devoured. At the time, cowbirds were known as buffalo birds.

The traditional theory has been that cowbirds were forced to adopt the habit of laying their eggs in other birds' nests to be free to move on with the nomadic bison herds. But today that theory's no longer certain. When the bison were almost exterminated at the end of the nineteenth century, cowbirds didn't miss a beat. They switched to the much more sedentary cattle herds that replaced the bison. Yet, the cowbirds didn't embrace nest building. In fact, we humans made life even easier for them by dramatically cutting back forests, allowing them to penetrate deeper into the woods and find new species of birds with nests to target.

Why would any species of bird allow a cowbird to lay an egg in its nest? Why doesn't the host bird just remove or destroy the intruding egg? It's not straightforward.

What *is* surprising is that many host birds don't take action. It's possible that they don't recognize that the cowbird egg is different, even though, unlike some nest parasites in Europe, cowbirds don't lay eggs that resemble those of their hosts. In a way, this isn't surprising because cowbirds are known to co-opt the nests of more than two hundred species! They couldn't possibly match their eggs to all. Maybe host birds aren't even aware of the number of eggs that should be in the nest.

And even if the host knew immediately that there was a foreign egg in its nest, it's not always clear what the bird should do next.

Getting rid of the cowbird egg seems like the obvious answer—and some birds do just that. Robins, cedar waxwings, and blue jays, for example, either puncture the shell or remove the offending egg from the nest, and for these birds, this is a successful strategy.

But fighting back doesn't always work. For instance, yellow warblers are prime targets for cowbirds. If a yellow warbler spots a cowbird egg in its nest, it will simply build a second nest on top of the first and lay a new clutch of eggs. This seems like an extreme response, but in fact, it's more efficient than building a whole new nest somewhere else. It's not always a great strategy, however; researchers have discovered as many as four warbler nests stacked one on top of the other—proof that the female cowbird, which lays about forty eggs in a single season, keeps watch and will return to lay another egg if her first (or second, or third) is abandoned. It might actually be better for the warbler to accept the cowbird nestling, even though it will rapidly grow bigger and hog most of the food the warbler brings back to the nest. Still, successfully raising just two young instead of three or four is better than running out of time building nests and raising none.

Deluxe
lofts
available.

Sometimes the cowbird female does more to ensure her progeny than simply laying a second or third round of eggs. In a long-term study of the prothonotary warbler, a bird that always accepts cowbird eggs, researchers uncovered two more aggressive cowbird strategies, which they called "Mafia" and "farming" behaviors.

The study site was a set of hundreds of nesting boxes for warblers—an astonishing 60 percent of which had cowbird eggs in them. When the scientists intervened and removed the unwanted eggs, the cowbirds returned and destroyed large numbers of those nests. This is the Mafia strategy: do what we want or we will destroy your home and kill your babies. The warblers' response was of course to start all over again, giving the cowbirds a second chance to lay their eggs.

"Farming" sounds more benign, but it's not. If the cowbirds had somehow overlooked a nest and laid no eggs in it, they'd return and destroy that nest, forcing the warblers to rebuild. That gave the cowbirds the chance to lay an egg, making up for overlooking the nest the first time around.

This study confirmed that cowbird females don't simply sneak in, lay an egg, and never return. They monitor what goes on in the nests they've taken over and retaliate if their offspring are threatened or lost. This might explain why so many species accept cowbird eggs and even adapt to their presence with strategies of their own. For instance, song sparrow young with a cowbird in their nest will imitate the volume and pitch of its begging calls, but they don't call as often until the food arrives. The scientists who oversaw this study speculate that the song sparrows let the cowbird do most of the calling to make the parents gather food at top speed, but they step up their own calls when the food arrives to ensure they receive as much as the cowbird does.

With forty eggs in a summer and more than two hundred species to target, you'd think cowbirds would have overrun North America by now. But while the organization Partners in Flight estimates a substantial breeding population of 120 million birds, that's actually a decline of 31 percent since the mid-1960s. Of course, most North American birds have suffered similar declines over that time, so it might simply be that today's cowbirds just have fewer nests to target.

Why were animals once so big?

It's true that there are many supersize animals in the world today: whales in the oceans; elephants, rhinos, and hippos on land. The blue whale holds the title of largest animal ever to have lived—its tongue alone can weigh as much as an elephant! If we're talking just about land animals, the elephant is the biggest today, but it's no better than a middleweight compared to the dinosaurs. And while ostriches are pretty big, there are no flying birds today that rival the pterosaurs of the past.

So, yes, there's been a lot of shrinkage over the past hundreds of millions of years, but the most intriguing changes have happened not among the giants, but among much smaller creatures, such as dragonflies and millipedes.

You just gotta put in the leg work.

Imagine you're walking beside a wetland 300 million years ago and a dragonfly cruises by, hunting for insect prey, just as its descendants do today. But this much earlier dragonfly has a wingspan of 70 centimeters (28 inches), which means it could barely make it through a modern doorway. Admittedly, that's the biggest set of dragonfly wings found so far, but a wingspan even half that—and there were many species that big—is off the charts compared to what we see today.

Think that's big? Ancient millipedes, the many-legged arthropods (they have more like a hundred legs, though, not the thousand their name suggests), were as long as or longer than an adult human is tall. We know this because we have fossil imprints of their bodies and their trackways, and those lines of little footprints are 50 centimeters (20 inches) apart. That is one wide-body millipede.

These are just a few examples of how different life on earth was 300 million years ago. But why have insects like dragonflies and millipedes shrunk to the sizes we see today?

Science Fact! *What's the biggest insect on earth today? That depends on what you measure: length, wingspan, or weight. A stick insect found in China is definitely the longest. It's 62.4 centimeters (24.5 inches) with legs stretched out— that's about the size of a small dog!*

Wingspan? The white witch moth, Thysania agrippina, has a wingspan of up to 28 centimeters (11 inches).

If weight is the criterion, beetles hold the title—although it's not clear which ones because reliable records are rare. But names like the elephant beetle, the goliath beetle, and the titan beetle (Titanus giganteus) give us a clue. All these beetles can weigh about as much as a chicken egg, but Titanus is my favorite. It can cut a pencil in half with a single bite and is apt to turn on any advancing entomologist and attack.

The most intriguing explanation is that the earth's atmosphere was very different back then. Today the air we breathe is 78 percent nitrogen and 21 percent oxygen, with traces of other gases. But 300 million years ago, the levels of oxygen were much higher, somewhere between 30 and 35 percent. (That's about the upper limit—any higher, and vegetation would burst into flames with the tiniest spark.) These high levels of oxygen might have allowed much-larger insects because of the way they breathe.

Rather than lungs, insects breathe using holes on the surface of their bodies that connect to tracheal tubes. These tubes transport air from the outside environment to the insect's interior, where oxygen feeds the tissues and keeps the organs functioning. This is obviously a pretty workable system, given that there are millions of insect species, but it has its limits. The bigger an insect's body, the greater its volume and the longer the tracheal tubes need to be to deliver oxygen to the deepest tissues. Flying insects like dragonflies have an especially high oxygen demand when they're airborne. But if insects become too large, the tracheal tubes would begin to fill the entire body, leaving little room for anything else. That creates a hard upper limit on size.

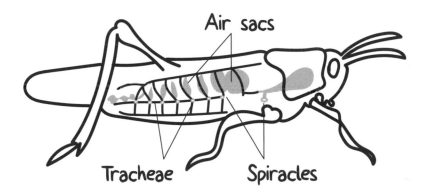

An experiment with beetles raised in the lab has backed this up. It demonstrated that the bigger the species, the more body space devoted to tracheal tubes, especially where the legs join the body (suggesting that this might be the crucial size-limiting bottleneck). The biggest beetles we know might have 90 percent of that area filled with tracheal tubes.

That brings us back to the oxygen-rich air of 300 million years ago. An atmosphere with higher oxygen levels would allow larger insects to circumvent this limit because more oxygen in the air is equivalent to more tubes in the body. An insect living in an oxygen-rich environment could grow larger without needing an excess of tracheal tubes. Some experiments have backed this up by showing that when dragonflies are raised in higher-than-normal levels of oxygen, they can grow to be 15 percent bigger.

These experiments suggest that oxygen might have been a critical factor, but we can't be sure it's the only one. For instance, there are relatively few insect fossils from the time, so we can't know how widespread the possible oxygen effect might have been. Maybe there were hordes of small insects, too. And the eventual decline in large insects was apparently part of a general reduction in all insect species—it would be difficult to pin all of that on oxygen.

Also, insect size doesn't correlate perfectly with peaks and valleys in ancient oxygen levels. According to the fossil record, there were two times in the geological past when, even though oxygen was stable, giant-insect populations shrank. The first of these happened when insect-eating birds came on the scene, and the second when bats began to proliferate. For both predators, one giant dragonfly is surely an easier meal than several smaller ones. This supports the idea that extra-large insects might have been able to attain their unique size only when they were free of predators. Once the hunter became the hunted, size was no longer an advantage.

You're full of hot air.

Why do cats' eyes shine?

SHINE A FLASHLIGHT IN A CAT'S EYES and the eyes shine back, usually with a yellowish or greenish hue. When light shines in a cat's eye, it first hits the retina, where much of it is absorbed. Some gets through, though, and immediately encounters a second barrier, called the *tapetum lucidum* (from the Latin for "shining layer"). Why is that there?

I've only got eyes for you.

The retina, the first tissue that light reaches, covers the back of the eye. It contains photo-receptors that capture light and trigger nerve impulses to the brain, where those impulses create vision. It's obviously crucial for good eyesight to trap as much of the incoming light as possible. But there's always some light leaking through the retina, and that's when the tapetum comes into play. It acts like a mirror, bouncing the light back and giving the retina get a second chance to absorb it. Even so, some escapes the eye completely—that's what accounts for the cat's eye-shine.

An astonishingly wide range of animals in addition to cats have a tapetum, including common fish like carp, various butterflies and moths, the American alligator, dogs, cows, horses, dol-phins, and even the Tasmanian devil. The sheer variety indicates that the tapetum is not a recent evolutionary development, and some scientists argue it first appeared among the earliest fish, more than 300 million years ago.

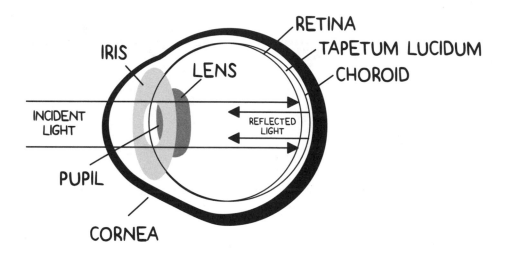

The modern descendants of those early fish—sharks, sturgeons, and the "living fossil" known as the coelacanth—all have tapeta. The coelacanth's is notable because it reflects greenish-yellow light, the wavelength of light that penetrates best to the depth of the ocean where the coelacanth lives. This seems to suggest that the tapetum of each species has evolved to func-tion precisely according to its needs. The tapeta also have different molecular structures, which accounts for the varying colors of eyeshine among different species. Evolution has allowed for many variations on the theme.

Did You Know . . . The eyeshine of cats and other animals is not the same as the red-eye effect that photos sometimes capture in humans. We don't have a tapetum, but we do have numerous blood vessels in our retinas and the underlying tissue. Light from a camera flash reflects off that tissue just as it does a tapetum, and that creates the red-eye effect—but in this case, the color is literally bloodred.

To avoid the red-eye effect when you take pictures, you can ask your subjects to look slightly away from the camera so the flash doesn't flood their eyes with light. If you're indoors, you can also increase the lighting in the room so that people's pupils aren't as dilated when the flash goes off. That's exactly what most modern cameras do by emitting a couple of quick flashes of light to constrict pupils before the actual picture is taken.

The tapetum is not the only unusual feature of a cat's eyes. The pupil is also a narrow vertical slit, not round like ours, and that improves visual acuity, especially at night and when hunting. Cats are ambush hunters—they wait for their prey—and they also hunt in both daylight and dark. Their eyes must be able to adjust to a wide range of light levels, and an eye with a vertical slit does that extremely well. A slit pupil can dilate ten or fifteen times as much as our round pupils can.

Cats also hunt close to the ground, and the vertical pupils are better for seeing details at the horizon from that height. Combine that with binocular vision, and a cat's eye is perfectly adapted to accurately track prey running along the ground. It's a mouse hunter's eye. Snakes hunt the same way, and their pupils are also vertical slits, but it's not that cats and snakes share these eyes because they're close relatives. It's something about the eye itself.

But vertical pupils aren't found across the entire cat family: lions and tigers have eyes with round pupils, like ours. Because of this, they lose a little of the ability to adjust to large variations in light, and their height above the ground deprives them of the ground-hugging house cat's advantage of detailed vision at the horizon. These big cats are also not ambush predators like the domestic cat; they run down their prey instead, and apparently round pupils are fine for that.

In the same study that revealed the advantages of vertical-slit pupils, American researchers showed that the majority of prey animals have horizontal-slit pupils. They argue this gives animals like cows sharper vision when looking out toward the horizon, which is important for both detecting the approach of predators and knowing in which direction to flee. When a grazing cow lowers its head to the ground, the eyes roll up in the sockets, allowing the animal to keep scanning the horizon.

Of course, none of this is ever as clear-cut as we'd like. Cows with horizontal pupils see the horizon better. So do cats with vertical pupils. Prey animals tend to have horizontal slits, but so does the ferret—a predator if there ever was one. Unfortunately, this large American study—214 species' worth—did not consider the ferret. But owners of ferrets know that they are extraordinarily interested in bouncing balls, which some scientists have claimed is because they're used to prey (like rabbits) that bounce. We need a new study or ten to understand how this might connect to the horizontal slit of their pupils.

It's remarkable that something apparently so simple—the shape of the pupil—reflects a complex mix of behavior, evolution, and optics across hundreds of species and thousands of years. And it's all there in your cat's eye!

What you are you looking at?

What is the fastest animal on earth?

THE TWO BEST-KNOWN EXAMPLES of high-speed creatures are the cheetah and the peregrine falcon. And they are indeed fast. There's some question about the maximum speed of a cheetah, though. The highest rate ever recorded was 112 kilometers (70 miles) per hour, standard highway speed for a car. But it's not clear how accurate this is. Most modern records suggest that cheetahs top out at around 100 kilometers (62 miles) an hour and are usually running even slower. That's barely fast enough to catch their antelope prey, but they do have two additional advantages: they can turn on a dime, and their acceleration is world-class. They can reach 75 kilometers (47 miles) an hour in just two seconds—comparable to a Bugatti Chiron supercar.

Do you think I'd cheat at this?

A cheetah is built much like a greyhound, another famously speedy animal, but the greyhound is capable of only about 70 kilometers (43 miles) per hour. A team at the University of Nottingham School of Veterinary Medicine and Science compared high-level video of greyhounds and cheetahs and showed that as the cheetah accelerates, it increases both the frequency and the length of its strides. The greyhound boosts the length only. Also, at its highest speeds, the cheetah shifts as much as 70 percent of its weight to its back legs, perhaps because as it nears its prey, it often strikes out with one of its forelegs and so relies on its hind legs for balance. Both these animals are, amazingly, completely airborne twice in the same stride: once when the hind feet pass the forefeet, and again when all four legs are stretched out. The greyhound breed is said to be four thousand years old, which sounds ancient but is still not a long time to develop these skills. Cheetahs have been around much longer.

The peregrine falcon is also a champion. When it's in full dive, it has been clocked at an unbelievable 389 kilometers (242 miles) per hour, faster than a bullet train and certifying it as the fastest animal of any kind in the world. But peregrines are much slower in level flight, so if you want to argue that using gravity is an unfair advantage, then your record holder is the white-throated needletail swift, at 169 kilometers (105 miles) per hour flying horizontally.

The swift has an unlikely challenger in the Brazilian free-tailed bat, which has been clocked at 160 kilometers (99 miles) per hour. The free-tailed bat is a surprising contender because generally bats are not as fast as birds. But this particular animal has unusually long wings, and aerodynamically that's a plus. On a really good day, it could surpass the swift. In fact, when you consider that the bats were clocked carrying a measuring device that was about 4 percent of their body weight, it's possible that they could go even faster when flying unencumbered.

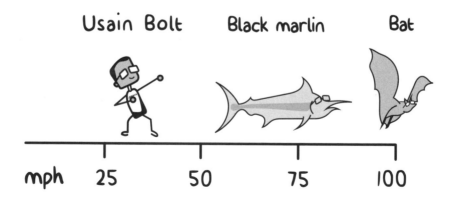

As a group, birds have the largest number of fast movers. Racing pigeons have hit about 150 kilometers (93 miles) per hour, and there's even a hummingbird that gets up to 100 kilometers (62 miles) an hour. Mammals are much slower, but even among the mammals, we humans are pretty unspectacular. Sprinter Usain Bolt, in his world-record 100-meter run of 9.58 seconds, reached a top speed of just 44 kilometers (27 miles) per hour. (That's still faster than a wombat, though.)

Fish are at a disadvantage because water creates friction, but the black marlin is generally considered the fastest fish in the ocean. Its top speed of 132 kilometers (82 miles) an hour is pretty remarkable, but maybe questionable: that speed was recorded by timing how fast a marlin on a hook stripped the fishing line off the reel. Lots of room for error there.

Science Fiction!   *The wildest claim ever made about an animal's speed came from an entomologist named Charles Townsend in the mid-1920s. Writing in the* Journal of the New York Entomological Society, *he reported that he had seen a deer botfly whip past at a speed he estimated to be 1,287 kilometers (800 miles) per hour. In other words, faster than the speed of sound. Physicist Irving Langmuir destroyed Townsend's claim by calculating that to move that fast, the fly would have to consume one and a half times its own weight in fuel every second, and air pressure would crush its head. He estimated its actual speed to be a much more pedestrian 40 kilometers (25 miles) per hour.*

These are the records for creatures whose speeds we can relate to best. But there actually is a completely different way of measuring speed, and that's by calculating the number of body lengths per second. This opens the door to some less well-known competitors. Let's start with Usain Bolt. His top speed of about 44 kilometers (27 miles) per hour translates to 6.2 body lengths per second. A cheetah reaches around 16. If Bolt could manage 16 body lengths per second, he'd be running more than 2,000 kilometers (1,240 miles) per hour. Amazing, but by no means championship performance.

The fastest animal on earth, when measured in body lengths per second, is a mite the size of a sesame seed. Mites are related to ticks and spiders, and this particular one, *Paratarsotomus mac-ropalpis*, can move at 322 body lengths per second, or more than fifty times the speed of Bolt. This crushes the old record, set by the Australian tiger beetle (which, at 171 body lengths per second, is still the world's fastest *insect*). Physicists think the *Paratarsotomus* mite is pushing the upper limits of what's physically possible. Why the need for such speed? No one knows because the mite is so little studied.

Physics is central to all of this. Two scientists in Paris have concluded that all living creatures, except those that fly, move at roughly the same speed when compared by body length per second. Everything from bacteria to whales (and that encompasses most life on Earth) moves at about ten body lengths per second—actually within a factor of ten around that number. Usain Bolt is a little sluggish at 6.2, and the cheetah is a speed demon at more than 15. The mite is still an outlier, but just about everything else has that maximum speed in common.

Why do some animals' eyes face forward while those of others face to the side?

THERE IS A STRAIGHTFORWARD ANSWER to this question, but as usual, "straightforward" means the answer is incomplete. Put simply, animals that are prey, like rabbits, zebras, sheep, and antelope, have eyes on the sides of their heads to allow them to see more of their surroundings—a kind of early-warning system for approaching predators.

By contrast, predatory animals like lions, tigers, wolves, and eagles have front-facing eyes, and this allows them to see in three dimensions, so-called binocular vision. Such vision makes it possible to tell exactly how far away something is. If an animal is going to pounce on its prey, it has to know exactly how far to pounce.

Did You Know . . . Some birds can see completely around them without moving their eyes or heads at all. These include the mallard duck and the American woodcock, which is also called the bogsucker, timberdoodle, hookumpake, wafflebird, worm sabre, and more—making it the bird with the most nicknames as well.

But as I said, this answer is incomplete. There are exceptions to this simple predator-prey classification. Among predators, mongooses and tree shrews on land and orcas in the ocean all have eyes on the sides of their heads. It's not completely clear why these animals are exceptions to the rule, although orcas, for one, also use the clicks and whistles of echolocation to hunt, so they may not have to rely as much on their eyes alone.

At the same time, some animals with forward-facing eyes, like fruit bats and several primates, are not predators at all, but fruit eaters. They don't need their binocular vision to chase down prey—although being able to judge the exact distance to hanging fruit while flying or climbing a tree might still be a big advantage.

These exceptions can be explained, but in other cases, we don't know exactly what's going on, eye to eye, between predators and prey. A study of snowshoe hares in Canada provides a good example. Snowshoe hares are definitely prey—as one researcher put it, "Almost everything eats them." Lynx are their main predators, although there are least nine others, and the hare's life expectancy in the wild is only a year or two (versus ten in captivity).

TRY THIS AT HOME! Take any card from a deck, hold it out to the side at arm's length, and stare straight ahead. Then, keeping it at arm's length, slowly move the card forward without shifting your gaze, until you can identify it. You'll be amazed at how far forward you have to bring the card before you can tell what it is. A cow would do much better at this test than you!

Scientists at Lakehead University in Thunder Bay, Ontario, studied the feeding behavior of snowshoe hares in an environment combining open space with a sprinkling of red pine and thickets of aspen. Hares eat both aspen and pine, and the scientists wanted to know what strategies they'd use to avoid predators—or, as the scientists put it, to survive in a "landscape of fear."

Why can't I just eat at IHOP?

You might think the strategy would be simple: minimize the time in the open and maximize the time under cover. After all, if you feed in the open, you're likely to be seen, and if you take cover in the aspen, you may not be. But funnily enough, the hares were actually *less* vigilant in the open than when they were sheltered by vegetation. Even though they fed for shorter periods in the open, they checked their surroundings more often when they were close to cover and presumably safer.

It might be that life is a trade-off for the snowshoe hare: the aspens provide cover, but at the same time, according to scientists, those "dense and tangled branches diminish sight lines." Yes, a distant predator might not see the hare, but the hare might also overlook the predator—until it's too late. Those circumstances are flipped in the open: prey are seen more easily, but so are predators, giving the hare time to flee.

Mark Changizi is the director of human cognition at 2AI Labs. He thinks eye placement has to do with more than just the difference between predator and prey. He argues that binocular vision is the next best thing to X-ray vision, because the sense of depth it provides allows animals with forward-facing eyes to see better in complicated environments, such as the confusion of branches, twigs, and leaves in those aspen thickets in the snowshoe hare study. The monocular, two-dimensional vision afforded by side-facing eyes can't do that.

Eye can go the distance.

TRY THIS AT HOME! Mark Changizi suggests you do this simple exercise to see how the depth perception of binocular vision can mimic X-ray vision. First, hold up one finger in front of your face and close one eye. Your finger will completely obscure whatever's in line with it across the room. Now switch eyes. The location of your finger will seem to shift (your eyes have different vantage points), and it will obscure something different. Finally, open both eyes. At last you can see the entire room and your finger, which now appears to be transparent.

Because snowshoe hares have eyes on the sides of their heads, Changizi would argue that visual clutter is a problem for them. When sheltered by visually confusing vegetation, they might be giving up as much as they're gaining. That could be why they remain as vigilant when sheltered by the trees as they are when snacking on pines out in the open.

If Changizi is right, the position of the eyes is determined not so much by whether an animal is predator or prey, but rather by where it lives. In leafy environments, 3D vision might be best, while in the open, it's more valuable to see on both sides.

Do animals mourn their dead?

WE CAN'T REALLY BE CERTAIN ABOUT THIS. Very different species *act like* they know when another animal is dead. There are heartbreaking stories of mother orcas, elephants, and chimpanzees keeping dead infants with them for weeks. But we don't know what those animals are thinking. Are they feeling something we would call sorrow? Or are we misreading their actions? They could have reasons we can't even imagine.

In most cases, we can't even be sure how animals tell when a member of their species has died. Ants, though, are an exception. Studies have shown that they use chemical cues—smells or tastes—to determine the presence of death. As a dead ant begins to decompose, chemicals called necromones are released. Worker ants sense these chemicals and react by dumping the dead

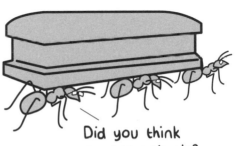

Did you think
we were animals?

35

insect with other garbage or burying it underground. The ants handling the disposal are known as undertakers, and they're efficient: even a live, kicking ant that's been smeared with oleic acid, a necromone also found in almost all cooking oils, will be carted off to the ant cemetery.

Ants are likely not "thinking" about why they're doing these things (or if they are, we would never know it). But disposing of the dead is a good survival tactic for preventing the spread of disease in the colony.

Did You Know . . . Ants are usually not so kind toward rival ant species. Some will even surround their nests with the heads of ants they've killed, presumably as a warning to others. This grisly tableau has nothing to do with mourning.

There are other animals that use odor to determine death (in fact, the smell of a decomposing body has an effect on most species), but while ants are attracted by it, most others are repelled. Even sharks avoid the odor of a dead companion, as do most mice and their relatives, voles and shrews.

Rats, however, are more like ants: the odor of a dead rat will prompt others to bury it. If a live rat is placed in a cage with a dead one, the live rat will try to bury the corpse by pushing bedding material onto it with its front legs. The older the corpse, the more time the rat will spend covering it up. The chemicals alerting the rat are putrescine and cadaverine. We can smell these chemicals, too, and as the names suggest, they're disgusting. For ants and rats, then, burial or disposal tells us they know when one of their kind is dead.

But some animals' responses are more elaborate. Crows, for example, may react in different ways to the sight of a dead crow. Kaeli Swift and John Marzluff at the University of Washington have set up experiments to illustrate what can happen. In one, they put out food for local crows until the birds became accustomed to using that feeding station. Then Swift and Marzluff placed a (stuffed) dead crow on the ground, with a nearby human watching to see what the response would be.

When the crows first spotted the dead bird, they produced alarm calls to attract others, and before long a mob of crows had gathered. The mob gradually dispersed, but for days after, the birds would avoid the feeding station. It seemed as if they were alerting others to the significance and danger of the place. They also continued to harass the person who had been monitoring them, even weeks after the corpse had been removed. If the corpse was presented alongside a model of a common crow predator like a hawk, their reaction lasted even longer.

 **Did You Know . . .** The crows had no reaction to a dead pigeon or a squirrel, showing that they indeed recognized when a corpse belonged to their own species. They weren't so much mourning the dead as warning the living.

In these experiments, the crows kept their distance from the dead bird, probably because there was either a human or a model hawk nearby. But when Swift and Marzluff placed dead crows alone on the ground in areas where there were already nesting pairs, things were different. Occasionally, the crow that first spotted the corpse would approach, mount it, and then be joined by its mate—and both would, as Swift put it, "rip it into absolute shreds." But a model of an apparently living, albeit unfamiliar, crow didn't elicit that violent response. So the local birds weren't mistaking the dead crows for dangerous interlopers or combatants. They knew death when they saw it, but they responded with more aggression than fear. Again, it's hard to know what they're thinking. Swift even tells a story of crows dropping sticks on a corpse!

Was it murder?

The animals that show behavior closest to mourning are the most intelligent ones: elephants, orcas, and chimpanzees. Elephants will spend long periods of time inspecting and even touching dead relatives that have been reduced to bone. They seem especially interested in the tusks and the skull. Both chimpanzees and macaque monkeys will carry dead infants with them, sometimes for weeks. It's a powerful portrait of a mother's bond with her baby.

Orcas do the same thing. In the summer of 2018, a female whale whose code name is J35 (she's also known as Tahlequah) carried her dead calf on her head for more than two weeks and across 1,600 kilometers (1,000 miles) of ocean along the west coast of Canada and the United States. News reports claimed that she finally abandoned the calf because it was decomposing, but ultimately we can only guess why she let the baby go and rejoined her pod.

With increasing brain size comes behavior that looks more and more like mourning. But even though the last few decades have reminded us that animals of many kinds are much more intelligent than we ever gave them credit for, their actual thoughts are still impenetrable to us. It's tempting to think, though, that our own ideas about death have their evolutionary roots in behaviors like this.

Why did scientists equip beavers with parachutes?

WHEN YOU READ THIS QUESTION, was the first image that came to mind a beaver with goggles and a leather flying helmet, hitting the ground, doing a shoulder roll, and cutting the lines to the parachute? Sorry to say that's not exactly what happened, but there actually was a time when beavers in cages were parachuted into Idaho from the skies above.

Beavers have lived in North America much longer than any humans. The first relative arrived perhaps 30 million years ago, when about a third of the continent was underwater. Several different beaver species descended from it, the most spectacular being *Castoroides*, the giant beaver, which was enormous—somewhere between the size of a Saint Bernard and a black bear. Given that beavers are better at cutting down trees than anything else, you'd think a massive, water-dwelling animal with six-inch incisors would have decimated the forest, but apparently *Castoroides* spent its time eating underwater plants, not felling trees or building dams.

Geronimo!

Did You Know . . . One extinct beaver genus known as Palaeocastor had a habit of digging intricate burrows that geologists call the Devil's Corkscrew. These were spiral tubes that extended straight down into the ground and could be as much as 2 meters (6 feet) tall. They were discovered by ranchers in Nebraska in the mid-1800s, and geologists and paleontologists had never seen anything like them before. Eventually, a close examination of markings on the walls revealed that these gopher-size beavers had used their incisors to dig these bizarre tunnels.

Although the beaver family tree has had many branches, only two species exist today: *Castor canadensis*, the North American beaver, and *Castor fiber*, the Eurasian beaver. Unlike *Castoroides*, *Castor canadensis* is all about chomping down trees and constructing dams and lodges to live in. It's much more at home in the water than on land. But its activities do change landscapes—often dramatically.

Picture this: A creek runs through a dense forest. Beavers happen upon it and decide to make it their home. They cut down some small trees and use them to create a dam across the creek. A small pond forms, flooded trees die, and old plant species move out while new ones move in. Frogs, muskrats, herons, and countless others take up residence. In time, the beavers move on, the dam disintegrates, and the pond disappears—but in its wake, a meadow forms. In this way, the wetlands the beavers created become hot spots of biological activity.

But humans who own land next to the creek may find it difficult to appreciate the long-term environmental benefits of felling trees and creating ponds. This is exactly the challenge that faced the Fish and Game Department of the Idaho government back in 1948. New subdivisions were bringing people and beavers together with predictable results. So the government decided to trap and transplant beavers to the wilderness. The problem was that it really *was* wilderness—mountains, forests, and few roads—and the only way the Fish and Game people could move the beavers was to pack them into crates, put them on the back of mules or horses, and spend the next couple of days trekking into the wilds. None of animals—or the people—liked the process, and many beavers died en route.

The science of transplanting beavers—what time of year to do it, what age the animals should be, how many should be moved to any one place—was actually well worked out. The science of how to deliver them wasn't. But then someone (that person has never been publicly identified) had the idea of putting them in crates and dropping them from airplanes. A good idea? No one knew until it was tested.

Step one was to use dummy weights to test parachutes. The scientists eventually settled on a parachute with a capacity of 64 kilograms (140 pounds). At that weight, they could attach one crate holding two beavers to each parachute. But how to allow the animals to exit upon landing? One early idea was to make the ends of the crate from woven willow branches. Since beavers just love willow, they'd eventually chew their way out. This promising idea was rejected, however, when scientists realized that the beavers might start eating the willow right away, chomping their way out of the crate in midair.

There were a few stops and starts, but it wasn't long before they had a parachute-and-crate pairing that looked good. They knew the best altitude to drop the package was between about 150 and 250 meters (500 and 800 feet); that would give the parachute plenty of time to open. And they had figured out a cool way of opening the crate when it hit the ground. It opened like a suitcase: In the air, the taut parachute lines pulled it together. But on impact, the parachute lines would collapse and heavy-duty rubber straps inside the crate, their tension now unopposed, would force the crate open. Now all they had to do was try it with a living beaver.

According to game warden Elmo Heter, writing in *The Journal of Wildlife Management*, the brunt of the testing fell on the shoulders of a fourteen-year-old beaver they'd named Geronimo. Geronimo was parachuted again and again. As soon as he landed, he was collected and taken to do it once more. Not a great way for Geronimo to spend his time, but he eventually capitulated; whenever he saw the game wardens coming to get him, he'd voluntarily return to the crate.

It all paid off. Geronimo was selected for the first real parachute drop and apparently success-fully started a colony with the females that accompanied him. By late 1948, seventy-six beavers had been safely parachuted into the wild. (One died when it somehow escaped its crate and either fell or jumped from the plane.)

A year later, the Fish and Game Department reported that all the transplanted beavers were thriving—which, if true, was an amazing result in the ongoing, sometimes uncertain relation-ship between humans and beavers.

Did dinosaurs brood their eggs?

HOW COULD THAT POSSIBLY WORK? A giant dinosaur settling itself on a clutch of fragile eggs? No matter how "gentle" the dinosaur, the outcome wouldn't be good. And yet evidence shows that not only did many species of dinosaurs build nests and lay their eggs in them, but they also tended to those eggs. It's not clear whether they were actually incubating the eggs—that is, keeping them warm the way birds do—but they were *acting* just like birds.

How do we know so much about the way dinosaurs behaved 100 million years ago? We have fossils—and some of them tell an amazingly detailed story. In Mongolia a few decades ago, paleontologists discovered the fossil of an oviraptor, a small dinosaur a little over 1 meter (4 feet) in length from the top of its head to the tip of its tail. It likely had feathers, too. This particular oviraptor had

I'm feeling rather bird-brained.

died on its nest, spread-eagled to cover as many of its eggs as possible. The eggs were also remarkable—there were two layers in total, and many eggs appeared to have been laid in pairs.

Other fossilized nests have revealed that dinosaurs arranged their eggs in circles—like the trays of shrimp in supermarkets. The bigger the eggs (so presumably the bigger the dinosaur that laid them), the wider the empty space in the middle. This suggests that while some dinosaurs did sit on their eggs, the big ones might have made sure that the bulk of their body was in the empty center, so as not to squash the eggs or the young.

So the answer to the question is yes, dinosaurs did brood their eggs—but only some dinosaurs in some circumstances. Others took a completely different approach. For example, many of the huge sauropods, such as *Apatosaurus* and *Titanosaurus*, apparently laid their eggs and then covered them up and abandoned them, like turtles and crocodiles do today. How do we know this? More cool science. First, we know that turtles and crocodiles have eggs that are more porous. If exposed to the sun and the wind, these eggs would dry out, but because they're buried underground, that's not a concern. On the other hand, birds' eggs, which are exposed to the elements in nests or even on the ground, have tinier holes in their shells to prevent moisture from dissipating. Researchers at the University of Calgary were able to categorize dinosaurs into those that likely buried their eggs and those that didn't based on how porous the shells of their fossilized eggs were. Recently, some of the same researchers have shown that early on, dinosaur eggs were soft-shelled, like modern turtles. Only later did hard-shelled versions appear.

But that's not all. Some of the same researchers also showed, based on the development of embryonic dinosaur teeth, that adults had to incubate their eggs for as much as six months, leaving themselves vulnerable to predators or starvation. (We know nothing about whether a brooding dinosaur had a mate to offer protection or gather food.)

It's so nice to get away from the eggs for an afternoon.

There are other clues to dinosaur nesting behavior besides shell porosity and incubation time. Color, for instance, is another difference between modern birds' eggs and those of turtles, snakes, and lizards. Animals that bury their eggs produce shells with no color and no speckles—presumably because there's no need to camouflage them underground. Birds, of course, are different: colored eggs with spots and speckles testify to that.

You can't tell the color of a fossilized dinosaur egg just by looking at it—the colors don't survive the fossilization. But the chemicals responsible for those colors do. In 2018, an international team of scientists bounced laser beams off dinosaur eggs and was able to tell, by the changes in the returning light, what pigments were present on the surface of the shells. They already knew that there are two pigments, one red-brown and the other blue-green, that account for the range of colors seen in birds' eggs today. They found evidence of the same two pigments in dinosaur eggs, which were laid long before any birds appeared on Earth. So not only did some dinosaur species lay eggs and brood them, but they invented egg color and patterning as well.

Science fiction? Today, the color and pattern of birds' eggs camouflage them from animals or other birds that would steal and eat them. Presumably that's why dinosaur eggs that were laid on the ground were patterned too.

But it's not just predators; birds have to cope with nest parasites, like the cowbird, which lays its eggs in other birds' nests. Does that suggest there might have been dinosaurian nest parasites as well? Imagine it: a dinosaur nest out in the open, dozens of eggs nestled inside, and the female, spread out over the eggs. In the underbrush, a small nest-parasite dinosaur lurks, waiting for the female to leave. When she does, the lurker jumps into the nest, deposits a few eggs, and takes off. Or maybe the intruding dinosaur sometimes squeezes in beside the incubating parent and deposits its eggs immediately. Or the egg layer dashes in when the nest is momentarily vacant, consumes some of the host's eggs, and replaces them with its own.

This seat's taken!

We see all those behaviors in cowbirds and their relatives, but I'd be willing to bet that they were not the first to try them.

History Mystery

Were pigeons once trained to guide bombs?

Absolutely true. And the person who trained them was the renowned, though controversial, B. F. Skinner, one of the most prominent psychologists of the twentieth century. Skinner was famous for insisting that researchers ignore what might be going on in someone's brain and concentrate instead on shaping behavior through reward and punishment.

While on a train one day near the beginning of the Second World War, Skinner was apparently thinking about how to build a device that would guide bombs to their targets with precision. At the time, the American military was struggling to make its bombs more accurate. If a bomb was released close enough to the ground to ensure it would hit the target, the plane carrying it risked being shot down, but if it was released too high, it would likely miss.

As Skinner watched a flock of birds flying in tight formation outside the train window, he was suddenly struck by the thought that with their maneuverability and exceptional vision, birds could be used as bombsights. He was experienced with guiding and controlling pigeons from his lab work, and thus was born Project Pigeon.

Skinner first set about devising a suitable, if somewhat rudimentary, guidance system. The bird had to be immobilized, so Skinner fitted its head and neck through a hole cut in the big toe of a man's sock, lightly tied its wings together with a shoelace, and strapped the lace to a piece of wood. Then he trained the pigeon to keep its head pointing—and pecking—at a moving target on a screen. If the bird successfully tracked the target, it earned a reward of seeds.

PROJECT
PIGEON

I feel a headache coming on.

No amount of seeds is worth this.

The next step was to engineer the system to work not with socks and bits of wood but with an actual bomb. The US military had an explosive with movable wings that allowed it to correct its course while en route to a target, and Skinner wanted to adapt his guidance system so that when the target moved off-center and the pigeon shifted its pecking to follow it, the wings were activated to maintain the bomb's focus on the target.

Skinner designed it to work like this: The force of the pigeon's pecks would open valves on each side of the target screen. Those valves would increase the air pressure in a system connected to the wings on the bomb. If the bomb drifted off target, the pigeons adjusted their pecking, the valves opened, the wings moved, and the bomb shifted back on course.

Over time, Skinner trained a team of sixty-four pigeons, and no matter what was thrown at them—pistol noises, bright lights, high-g forces—they just kept pecking at their targets. Male pigeons weren't even distracted when females were strapped in beside them. When fully trained, a pigeon could peck at a target four times a second for two full minutes. Some could reach incredible numbers, like 10,000 pecks in forty-five minutes—more than enough to keep a bomb on target.

Skinner made some refinements along the way. For instance, the pigeons soon figured out that they'd get a reward even if they didn't hit the target exactly, so Skinner added two intersecting light beams to the machinery. To get any grain, the birds had to peck precisely enough to interrupt both beams. Soon, they were nearly perfect at keeping the target in the crosshairs.

And just in case an individual pigeon did get distracted, Skinner hit on the idea of a three-pigeon bombsight, where the majority ruled. So even if one pigeon lost track, the other two kept the bomb on target.

Science _Fact!_ *Skinner's pigeons were just one species involved in war plans. At one time or another, people thought about training dogs to guide torpedoes to submarines or to blow up tanks, using seagulls to track submarines from the air, recruiting dolphins to search out underwater mines, and releasing bats as incendiary devices.*

This last idea was taken seriously. In this devilish plan, bats carrying tiny time-delayed fire-starting devices would be released from airplanes over targeted areas at night. At about the time these nocturnal creatures would take shelter for the day, under eaves or in chimneys, the devices would light, starting multiple fires.

But the project was plagued by shortcomings. For example, the bats were cooled to put them into semi-hibernation for the airdrop, but many either woke up too late or too early. Also, the clips that attached the device to each bat's chest often didn't work, and sometimes the cardboard boxes holding the bats failed to open in midair as planned. Eventually, the scheme was abandoned, to the chagrin of at least one enthusiast, who claimed that setting thousands of simultaneous fires in Japan would have ended the war without dropping the atomic bomb.

As flawless as the pigeon system was, Skinner was unable to persuade the military to use it. It was just too "utterly fantastic," he said. Finally, after hemming and hawing over Project Pigeon for years, the military scrapped the plan, citing more pressing technological needs. Skinner was disappointed, but he decided to hang on to the pigeons to see if they would preserve their targeting skills. Even six years later, they still performed beautifully.

The ethics of sending pigeons to certain death were apparently never taken into account. In fact, Skinner dismissed such concerns, calling them "a peacetime luxury." Happily for the pigeons, they were replaced by advanced technology—something the military was more comfortable with.

Part 2
Bodily Brainteasers

Will we ever fly like birds?

THIS QUESTION IMPLIES MORE THAN IT SAYS. It should really be, Will we ever fly like birds by using only our own muscle power and flapping our arms like wings? There are ways we can already fly using only human power, but to do it by flapping wings is much, much more challenging.

Most of our best efforts don't satisfy these requirements. Three outstanding examples are the wingsuit, the jet suit, and the glider-like *Gossamer Albatross*. The wingsuit is just that: a suit that adds surface area to the body with material that stretches from the wrists to the ankles and in the space between the outstretched legs. Surface area is crucial because it increases lift and keeps the person wearing the wingsuit from plummeting to the ground. But soaring with a wingsuit is not flying—it's gliding the way a flying squirrel does. Both humans and squirrels have a glide ratio of about 2:1 (that is, for every two meters they move forward, they fall one meter). Wingsuits are sometimes called squirrel suits because of the resemblance.

I play to wing.

Unlike a bird, a wingsuiter can't take off, but must either leap from a cliff or jump from a plane and glide to earth, deploying a parachute to land. Even though it may look as if the human in the suit is flying because it seems so natural, it's still gliding, not flight. On the other hand, a jet suit *is* flying. Pilots with jetpacks strapped to their backs can take off from a standing start, have traveled up to speeds of 137 kilometers per hour (85 mph), are able to ascend and descend, and can even soft-land without a parachute. But the pilots are not providing the power. The world's fastest jet suit, designed and piloted by British inventor Richard Browning, uses five jet turbines, two fitted to each arm and one on the back. It is fantastically maneuverable and requires tremendous strength and balance, but Browning's contribution is limited to guiding, not powering.

There have been aircraft that flew under human power, and some of those were spectacular. One, inventor Paul MacCready's *Gossamer Albatross*, flew clear across the English Channel in 1979. To power it, a single human pilot pedaled to turn a propeller. The name of the aircraft hints at its structure: "gossamer" because the aircraft was almost as insubstantial as spider silk (it was made of ultralight materials wrapped in plastic film and weighed only 32 kilograms (71 pounds). And "albatross" because the wings were extra long to provide as much surface, and therefore lift, as possible.

Did You Know . . . According to Greek legend, the first person to try to fly like a bird was Icarus. He covered his arms with feathers coated with wax, but when he flew too close to the sun, the wax melted, the feathers dropped off, and Icarus fell to the sea and drowned.

The *Gossamer Albatross* came oh so close to flying like a bird, but those amazing wings didn't flap. There's only one aircraft that has achieved human-powered flight with flapping wings, and that was an ornithopter, a craft named for Leonardo da Vinci's 1485 flying machine.

Built at the University of Toronto, the ornithopter, called *Snowbird*, not only set world records but also showed how difficult it is to achieve human-powered flapping-wing flight—and how unlikely it is that we will ever do much better than the 19.3 seconds it managed to stay aloft.

What a flight of fancy.

Todd Reichert, an engineer at the University of Toronto's Institute for Aerospace Studies, piloted—and powered—the ornithopter during its record-breaking 145-meter (475-foot) flight. *Snowbird* weighed less than 45 kilograms (100 pounds), and Reichert himself shed 8 kilograms (18 pounds) in preparation for the flight. But as light as the aircraft was, it had an enormous wingspan, about the same as that of a Boeing 737. To power *Snowbird*, Reichert pumped his legs and a series of ropes and pulleys then pulled the wings down to create lift. Reichert had to generate somewhere between 700 and 800 pounds of force (3,100 newtons) for each of the sixteen wing flaps on the brief flight.

Science _Fact!_ *History is littered—literally—with failed attempts at human-powered flight. Most were hilariously naive, although the end was rarely hilarious. For example, in 1503 an Italian named Giovanni Danti attached wings to his arms and jumped from a tower in Perugia. He survived but was seriously injured. Four years later, John Damian, an Italian-born physician in the court of King James IV of Scotland, leaped from Stirling Castle, again with wings, and suffered a broken femur. And those two were the lucky ones—many others died.*

Why can't we just strap on some wings and take off? It's partly weight and partly technique. Birds have a couple of important advantages over us—and the biggest is that their breast muscles are enormous for their weight, while ours are not. Most of the force that flaps the wings is generated by the breast muscles. Our breastbone, the sternum, isn't large enough to anchor muscles of the size we'd need to flap our arms with the power to get us off the ground. As early as the 1600s, scientists were making this argument. One even calculated that we'd need shoulders 6 feet (2 meters) wide to accommodate the muscles needed to fly. (Twenty-first-century calculations have suggested that the wings would need to be nearly 23 feet, or 7 meters, across!)

A bird's breastbone is different. It has a large plate called a keel running down the middle, and that adds a lot more area for attaching muscles. Also, a bird's bones are different from ours. Many are hollow, which you might think would make the bird lighter and flight easier, but, in fact, they're also quite dense. Greater density lends greater strength. The hollow bones also have air sacs in them, extending the bird's lungs into its wings, and therefore providing more oxygen for the demands of flying. By comparison, we are just too heavy and too weak.

The truth is that inventors who've tried to flap wings made of wood were never going to fly. A bird's wing isn't rigid. In flight, it's constantly curving and twisting—up, then down—being pulled close to the body, then straightening out. All these movements happening together will alter the amount of lift or drag so the bird can control its speed and altitude. Anything is possible: the bird soars, dives, hovers, even lands. Feathers can spread apart or tuck together, controlling the flow of air between them. A bird's wing is an amazingly complex aerodynamic mechanism. Nothing we've built compares.

Flapping your wings and soaring above the trees is a wonderful dream. Maybe it will be possible in space colonies with reduced gravity. Just not here.

What will future humans look like?

How to begin to answer this question? It would be reasonable to start with what we know about how modern humans evolved from our ancestors. But we're still recovering and analyzing prehuman fossils, so there are serious limitations there. What's more, we can't predict the environments in which our descendants will live. Evolution, after all, is the interplay between living things and their environment. So predicting future humans is inevitably guesswork, but let's give it a go.

The picture painted of human evolution used to be straight-forward. A four-legged animal came down from the trees (likely because of a climatic change that reduced forest and created vast savannas in Africa), then graduated to being bipedal, or walking on two legs. This freed the creature's hands, which evolved into versatile, tool-using graspers. At the same time, the brain expanded, enabling it to better

Welcome to the future!

manipulate the hands or even to invent new things for them to do. And remember, all of this played out over millions of years.

That simple picture has been dramatically clouded by new fossil finds over the last several decades. For instance, we now know there were creatures whose brains remained relatively small even while they were spending much of their time on two legs. We also know that at any one time, there was likely more than one species evolving in a humanlike direction, even if some of these species were not our direct ancestors. Human evolution was definitely not a simple straight line to us.

What can be said with certainty is that the combination of brain, hands, and two-leggedness conspired to make us what we are today, albeit with many detours along the way. These physical developments were all driven by evolution, but experts disagree about the future of human evolution. There's not just the "what will the environment be like?" question, but some argue that given the ease with which populations mix today, there is no longer an opportunity for discrete human populations to evolve on their own, the way it happened hundreds of millennia ago. Also, survival traditionally determined which genetic types flourished, but today survival is influenced as much by medicine as our genes.

On the other hand, there's good evidence that the human genome continues to change, and in that sense evolution is steaming ahead. The tricky thing is that it's not yet clear what effect these ongoing genetic changes will have.

Will we change physically? The picture here is confusing, too, because several significant trends of the past hundred years, like increasing height, have slowed or even reversed in some countries. And because of the size of the birth canal, the trend toward a bigger brain has pretty much stopped. So there's no reason to think we will have giant eyes, bulbous heads, and tiny noses and mouths, like aliens in some Hollywood films.

Most theories about future humans are built around the possibility of introducing technology into the human body. This could take many forms. If genetic engineering continues to develop as quickly as it has to this point, then not only could humans be protected against a myriad of diseases and even aging, but facial features and bodies could be genetically tweaked to suit fashion. Parents could decide that their baby should have a short, straight nose and full lips, or whatever is popular hundreds or even thousands of years from now. (Anyone talking in terms of thousands of years should remember that 40,000 years ago Neanderthals were living in Europe. It's a long time.) This would be evolution built not on genes but gene technology.

Or imagine a world where actual micromachinery is routinely introduced into the human body to maintain health. Future humans might stay in good shape not with exercise and a healthy diet but with nanomachines eliminating incipient cancers and other diseases. Survival would simply require having a body that could support all these devices. Being able to hunt, run fast, or use tools wouldn't matter much.

The ultimate step might be to connect the human brain to a computer interface that could both enhance the power of that brain and link it to many others. At that point, the physical body could become irrelevant. Whether you think that's good or bad may depend on whom you listen to, or if that person's having an off day: philosopher Nick Bostrom has given a lot of thought to uploading human intellect to computers, and he can go either way: imagine either a new world of unimaginable mental complexity or a machine-driven melding of minds to create an efficient but uninteresting supermind.

Last word on this goes to theoretical neurobiologist Mark Changizi, who thinks that brain-machine interfaces are not what we need—and indeed are unlikely to happen anytime soon. Instead, Changizi says, we should rely on our existing brain's ability to use its brilliant software to come up with new ways of dealing with the world. He points out that three of the things our brains do very well—speaking, reading, and playing music—were not the products of evolution. Instead, our brains began with the capacity to recognize and interpret visual patterns in our surroundings and to analyze natural sounds like water, trees, and birdsong, then turned it into the ability to speak, read, and play music. Each of these three skills was fashioned using preexisting neural networks in the brain.

Given the uncertainty of where we're heading and how fast we might get there, it's not surprising that predictions about future humans can range so wildly, from people who might look pretty much like us to an unrecognizable humanity whose life exists only in electronic circuitry. It's like the old joke: The pilot gets on the intercom and says, "The bad news is that we're lost. The good news is that we're making excellent time."

Is your brain really necessary?

This question was posed at a scientific conference more than forty years ago. The man who proposed the question, a British pediatrician named John Lorber, was recounting some amazing cases he'd seen where people who seemed unremarkable in every way were in fact shown to be missing much—even most—of their brains. His question was a theatrical way of asking, "What's going on here?"

One of his prize examples was a young college mathematician with a high IQ who, in Lorber's words, had "virtually no brain." A scan revealed that instead of filling the skull, the student's brain was nothing more than a thin layer of tissue clinging to the inside of his cranium; the rest of the skull was full of cerebrospinal fluid.

If I only had a brain.

The student suffered from hydrocephalus, or water on the brain, a condition where fluid that normally flows throughout the brain is either overproduced or fails to drain properly. Many children with hydrocephalus can be cured by surgery, but in severe cases, the fluid can slowly accumulate, flooding the skull and compressing the brain. The math student, as the brain scan revealed, was left with what appeared to be only a small percentage of the original brain tissue. Yet, he seemed perfectly normal—highly intelligent even.

Lorber's paper was met with much skepticism, partly because brain scans at the time didn't have anywhere near the detailed resolution they have today, but also because scientists—then and now—have a hard time believing anyone could lose that much brain tissue and still flourish. And Lorber never published his ideas in the scientific literature.

But there have been other cases like the one he described. In 2007, a Frenchman went for a checkup because he was experiencing weakness in one leg. Doctors realized that he, too, had hydrocephalus, and the pressure on his brain was affecting the neural control of his leg. They were able to correct the problem, but images showed he was missing as much as 50 to 75 percent of his brain. And yet he had a job and a family, and he seemed to be living a routine life, although his IQ was just 75, well below average.

Ordinary people leading normal lives with little brain tissue. Had they not happened to visit doctors, no one would ever have known their brains were so unusual. There are likely more of them, living their lives with no idea their brains are not like everyone else's. But how would we know? You can't justify scanning an entire population to see if anyone has an unusually small brain. Such brains will be discovered either accidentally, in a medical examination, or postmortem if they are donated for medical research.

Which leaves unanswered the question of whether the brain is really necessary. It's true that brain size isn't the same as intelligence. Russian novelist Ivan Turgenev's brain weighed over 2,000 grams (more than 4 pounds), but writer Anatole France's was about half as big. Yet both men were smart and accomplished!

But of course we're dealing here with much-greater apparent discrepancies in brains. Unfortunately, none of these individuals have had their brains examined after death, so no one knows exactly how the neurons and their companion cells are organized. Could the brain successfully reorganize itself to cope with this kind of reduction?

My left brain knows what my right brain is doing.

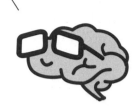

There are some hints from the treatment of epilepsy. In severe cases where seizures cannot be controlled by drugs, doctors sometimes remove the piece of the brain that's thought to be the source of the problem. This is a drastic step and is considered only when the patient is young and the brain is still developing. Young people are more likely to recover because of something called brain plasticity—when brain tissue is lost, new neurons will grow or neighboring areas of the brain will take over some tasks. Brain plasticity is not limited to young people—adults who've suffered strokes (where blood supply is interrupted and part of the brain is damaged) are often able to regain movement and speech—but generally speaking, children have better outcomes.

A recent case in Pittsburgh has shed light on what can happen in the aftermath of such dramatic surgery. A seven-year-old boy with untreatable epilepsy had a significant piece of his right brain removed. This included part of the visual cortex, which is responsible for vision, and another section for recognizing faces. Over the three years following the boy's surgery, scientists tracked his abilities in these areas, as well as his general mental development. His IQ scores were above average, proving he hadn't lost any general brainpower, but he did have a significant visual impairment—he was no longer able to see anything on his left side. This is because information from each eye travels to the visual cortex on the opposite side of the brain: your left eye messages your right visual cortex and vice versa. Because the boy had lost so much of his right visual cortex, the left side of his vision was a blank. He had to turn his head to the left to see anything on that side.

What was most exciting, though, was that the boy's ability to recognize faces, which is almost always centered in the middle of the right side of the brain, had been taken over by a similar area on the left side. Testing showed that a patch of neurons normally responsible for recognizing words had started to respond to the sight of faces. Word recognition was still intact, but somehow facial recognition had taken up residence in the same area. It was a beautiful demonstration of brain plasticity.

Does this case suggest anything about the more extreme losses of brain tissue due to hydro-cephalus? Perhaps. But without more detailed study of those cases, it's hard to imagine how simply reassigning neurons could possibly compensate for the loss of 75 percent or more of the brain.

A few neuroscientists argue that these extreme cases call into question the assumption that all our mental activity is based in the brain. How can that be, they ask, when a mostly empty skull can sometimes generate normal mental activity? This is all a great neurological mystery, and it's not clear when—or if—it will be solved.

I feel like I'm not all there.

What are lucid dreams?

LUCID DREAMING—A KIND OF DREAMING that allows you to take control and decide how a dream will unfold—must be one of the coolest things ever. For one thing, it shows that some parts of your brain can be awake while other parts are asleep. And it also proves that we can be conscious—that is, thinking and aware—in three different states: awake, dreaming, and lucid dreaming.

Does everyone dream? It's difficult to be sure of that, but it is true that everyone experiences what's called REM (rapid eye movement) sleep. It's called REM sleep because a person's eyes move rapidly back and forth under closed lids, and this is when most dreams take place. But there are people who, when wakened directly out of REM sleep, say they weren't

I must be dreaming.

dreaming. Does this mean that they actually weren't dreaming, or simply that the memory of those dreams disappeared almost instantly? It's hard to say, although there's no doubt that the memory of dreams *is* fleeting. While you might spend an hour and a half each night in REM sleep, you usually don't remember more than a few minutes' worth of dreams.

Science <u>Fact</u> or <u>Fiction</u>? Do the eye movements in REM sleep track the events of dreams? The answer to this is still controversial, partly because there are examples that don't fit. People who have been blind from birth, for instance, have rapid eye movements but don't have visual dreams. Even fetuses in the womb have their own version of REM, although they, too, are not seeing anything. But when sleepers are awakened and asked about their dreams, there's sometimes a correlation—perhaps their eyes moved up and down when they were dreaming of climbing a set of stairs. And while most of us are almost paralyzed in REM sleep, some people kick their legs and wave their arms, and those movements have been shown to match events in the dreams they were having.

It's difficult to generalize about dreams, but for many people, they're a little crazy. Strange events pile up, sometimes involving childhood places and people, sometimes not; some make a little sense, and some none at all. This is where lucid dreaming comes in. In lucid dreams, people know they're dreaming. They're actually able to say to themselves, "Wait, this isn't real. This is a dream." Not only that, but skilled lucid dreamers are able to take over and direct the action of the dream. No longer is the dreamer a helpless participant.

The best estimates suggest that slightly more than 50 percent of people have had at least one lucid dream in their lives. And those experiences can be extraordinary. The man who coined the term

This is a dream.

"lucid dream" in the early 1900s, Dutch physician Frederik van Eeden, described one of his many lucid dreams this way: "I dreamt that I was lying in the garden before the windows of my study, and saw the eyes of my dog through the glass pane. I was lying on my chest and observing the dog very keenly. At the same time, however, I knew with perfect certainty that I was dreaming and lying on my back in my bed. And then I resolved to wake up slowly and carefully and observe how my sensation of lying on my chest would change to the sensation of lying on my back. And so I did, slowly and deliberately, and . . . it is like the feeling of slipping from one body into another."

If you'd been standing at van Eeden's bedside, how could you have known he was having this fantastic lucid experience? One research approach has been to ask people to make a prearranged set of eye movements—say, eight left-right shifts—as soon as they're in a lucid dream. In a sleep lab, the eye signals can easily be picked out from the rapid-eye movements of the dream.

So how does lucid dreaming work? Brain wave recordings appear to show that the front of the brain has to partially wake up before a dream can become lucid. The front of the brain, the frontal lobes, is responsible for characteristics of wakefulness, like working (or short-term) memory, decision-making, and self-awareness. We all know that one of the dramatic differences between dreaming and being awake is that there's a kind of consistency and flow to waking life that is often absent in a dream. Dreams mimic waking, but only partly.

While most lucid dreamers do it for their own amusement, they're shedding light on consciousness at the same time. It's widely accepted that waking consciousness and dreaming consciousness are different—you simply have to compare the two experiences to realize how different they are. But lucid dreaming is a blend of the two and so qualifies as a third type of consciousness, one with some of the wildness of dreaming and of the rationality of being awake.

TRY THIS AT HOME! If you're interested in having lucid dreams, there are some techniques you can try. One involves getting yourself used to testing the reality of the situation you're in. This can be as simple as asking yourself throughout the day, "Am I dreaming?" Or if you wake up in the middle of the night, force yourself to stay awake for ten minutes or so, rather than trying to fall immediately back to sleep. Supposedly this allows you to drop right into REM sleep, improving your chances of dreaming lucidly. If all else fails, you could just rehearse a phrase like "The next time I'm dreaming, I'm going to realize it's a dream." None of these techniques are guaranteed to work, but they've all been shown to help some people.

Science Fact! *Sleepwalking is not related to dreaming. In fact, sleepwalking and sleep talking (known as somnambulism and somniloquy) occur in the deepest stage of sleep, and the affected person usually can't remember thinking or experiencing anything—even if she was busy opening doors, putting on her coat, or driving a car!*

Is muscle memory a real thing?

Most of us think muscle memory is the reason we never forget how to ride a bicycle: our muscles just kick into action and "remember" how to balance, how to lean into a turn, how to pedal faster to stay upright. While it's true that the muscles do the work, it's not actually the muscles that remember—it's the brain.

No muscle does anything without first receiving a signal from the brain. Every muscle cell in your body is supplied with a nerve cell, a neuron, that signals the muscle to move when it

Don't forget to
bring the muscle.

should. In turn, all of these nerves trace their origin back to the brain. The part of the brain responsible for triggering the nerve signals is called the motor cortex. This strip of neural tissue across the top of the brain is actually laid out as a map of the human body, with the legs and torso at the top, the hands and fingers partway down, and the face, especially the lips and tongue, on the side. (The lips and tongue are exaggerated in part because they're responsible for the complex movements associated with speech.) The map on the left side of the brain controls the right side of the body and vice versa. As you've probably guessed, the part of the cortex responsible for arms and legs would be especially involved in learning to ride a bike and preserving that memory.

So if it's true that you never forget how to ride a bike, it's your brain you should thank for that, not your muscles. When you were first learning to ride, and until your brain was fully trained on that skill, you probably experienced a few painful falls. The circuits of neurons responsible for the balance and coordination of bike riding were just being set up. Their numbers were growing, and more secure connections were being made and unnecessary ones were being deleted. It was a dynamic time for your brain.

Did You Know . . . A study of string musicians (violinists, cellists, and guitar players) showed that larger areas of their brains are devoted to sensations from the four fingers on the left hand—in other words, the fingers that press on the strings to play a note—than to those from the thumb, which just supports the neck of the instrument. But all of the left-hand digits occupy more brain space than their counterparts on the right. The size of brain area correlated with the age the musicians began playing—the younger they started, the more their brains changed. This study was different because it looked at touch information arriving at the brain, rather than outgoing movement messages. But it would be a surprise if the players' brains weren't set up for complex finger movements, too. Curiously, some scientists who've studied preserved tissue from Albert Einstein's brain have claimed to see a specialized "knob" of tissue that reflected the hand movements of his violin playing.

Some of the most intriguing research in this area has involved teaching people to juggle—not circus-worthy juggling—but just keeping three objects in the air, the classic three-ball cascade. Juggling is a nice example of hand-eye coordination: But what does that mean? Where in the brain does that develop?

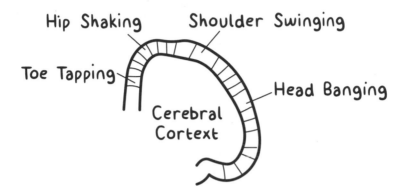

In one study, volunteers were divided into two groups—those in the first group would be trained to juggle, while those in the second group would not. Over three months, the trainee jugglers were coached and given tips, and they became adept at keeping the balls in the air for at least a minute. Using MRIs, researchers imaged the jugglers' brains before the training began and after the three months were up. A third scan was performed after they had refrained from juggling for another three months.

The first two scans recorded an expansion of gray matter (neurons and their extensions) from the beginning up to the peak of training. But the third scan showed that when the jugglers stopped practicing, their skills declined *and* their gray matter shrank. (Those in the nonjuggling group showed no changes in all three scans.) The increases in gray matter happened not in the brain's motor cortex, as you might expect, but closer to the back of the head, where vision is processed. This prompted the scientists to claim that the changes upgraded the "storage and processing of complex visual information." Fair enough—when juggling the eyes must guide the hands.

A second study looked at twenty-four novice jugglers who trained for six weeks. That study revealed that white matter, which is a kind of insulation wrapped around long extensions of neurons, had also increased in specific areas of the jugglers' brains, and it was still above initial levels even four weeks after the volunteers had abandoned juggling. This points to a concrete brain response to the practiced actions of juggling. But unlike the first study, this one showed no correlation between juggling skill and brain changes. The brains changed according to the total training time, not to how skilled any one volunteer became.

Juggling isn't bike riding, but it makes sense that similar changes would be happening in the brain, although with cycling the brain would be developing its ability to balance, sometimes at high speed, rather than trying to coordinate hands and eyes. What these juggling studies, which ran for only a few weeks or months, don't reflect is the lifelong learning of bike riding. If you haven't ridden a bike for years, you can still ride perfectly well, even though you might be a bit shaky at first. That suggests that some of the effects on neurons are permanent, but some need to be refreshed.

Experiments have shown that when people become so adept at a task that it's automatic, they can take on another task simultaneously, using the same areas of the brain for both (yes, you can walk and chew gum at the same time). The so-called muscle memory for bike riding, once established, should allow you to, say, talk to someone who's riding with you, and of course people do that all the time. Some experiments have shown that the brain doesn't have to work as hard at something that's become automatic. That might leave plenty of neurons free to tackle something else.

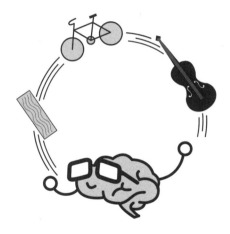

Did You Know . . . One of the most amazing cases of learning without thinking is that of Henry Molaison, a man whose brain surgery accidentally left him incapable of forming any new conscious memories. If you'd asked him what he had for lunch that day, he wouldn't have been able to tell you. But he was able to learn and become quite skilled at tricky tasks of manual dexterity, such as tracing a path using a mirror, even though he couldn't remember ever having done them before.

Why do we see stars when we bump our heads?

HAS THIS EVER HAPPENED TO YOU? You back into an open cupboard door and hit your head hard enough that points of light float in front of your eyes. You're "seeing stars." They might last for a few seconds or maybe even a minute. Then they fade.

These so-called stars are created by your brain. They're an invention of its visual system. When you hit your head, your brain is jarred, and either that motion or the impact of brain on skull triggers nerve activity in your visual cortex at the back of the brain. You create your own images.

I'm star struck.

TRY THIS AT HOME! If you want to see stars without hitting your head, there is an easier way. Close your eyes and press very gently on them with the tips of your fingers. After a few moments, points of light, swatches of color, circles, even elaborate geometric arrays, will all appear before your eyes. In this case, pressing on your eyes stimulates the retinas, which then talk directly to the visual cortex. Pressing on different places on the eye can even create different images. Sometimes you can get the same effect simply by keeping your eyes closed in a dark room. People held captive for long periods in dark cells often experience this phenomenon, which is why it's sometimes referred to as prisoner's cinema.

These lights are phosphenes, and they can be created in different ways. That blow to the head is one way (by directly impacting the visual cortex). But pressing on the eyeballs is a little more like actual vision. Neurons in the retina have one job: report light. Those nerve signals travel from the front of your brain to the back, the visual cortex, where they're rendered as the incredible variety of shapes and colors we see.

Phosphenes can also be triggered by stimulating the brain electrically. Experiments with human subjects have shown that as the frequency of such electrical impulses rose, the phosphene patterns changed, but when they reached 40 cycles per second, everything suddenly went dark. Every phosphene disappeared. Biophysicist Gerald Oster described it as "a feeling of being left alone in space."

Phosphenes are at the center of two unusual and very different stories. The first takes us tens of thousands of years into the past, to ice age artists painting cave walls in France, Spain, and other countries. Those walls are covered with fantastic images of leaping and running animals, handprints, and enigmatic geometric patterns of circles, dots, and spirals. Some archaeologists have argued that because the latter are so similar to the phosphenes that we can experience, they must be the result of prehistoric artists experiencing the same thing. The most common suggestion is that the artists were shamans of some kind who induced trances with psychedelic drugs, which can create phosphene patterns. Perhaps as the artists were in these trances, they would simply trace what they were seeing on the cave walls.

Phosphenes are also part of a modern story that involves stimulating the brain to create visual images. This research began a century ago, with neurologists applying electrodes to the visual cortex of people with brain injuries, some of them open wounds, to see what effect a weak electrical current would have. One patient reported seeing irregular rings and stars. This showed that the eyes aren't needed to produce visual effects—vision is created at the back of the brain. These experiments gradually led to the hope that stimulating the visual areas of the brain could restore sight in the blind.

Must have been a good trip.

Unfortunately, nearly a century has passed without producing any well-established way of doing this. It is an enormously difficult challenge that involves implanting electrodes in the brain in the hopes they will somehow mimic the activity of millions of neurons in the visual system. The complexity of the human eye-to-brain connection is daunting.

However, scientists have tried to take advantage of the fact that the visual brain generates an accurate map of what the eyes are seeing. It's accurate to the smallest detail, like the arrangement of lines to represent single letters of the

I feel we have a connection.

alphabet. But early attempts to use this map to bypass the eyes and spark phosphenes arranged like letters failed. Even though each electrical signal did generate an individual flash of light, triggering several at once resulted in an unreadable smear of phosphenes, not the straight lines and curves of a letter.

This illustrated how tricky the problem is. In the visual system, some neurons fire in response to horizontal lines and others to verticals. So the letter *T*, for instance, activates both. But neurons that are electrically stimulated respond to all different angles—resulting in a blob of light.

Recently, however, Michael Beauchamp, the director of the neurosurgery lab at Baylor University in Texas, has improved the technique. Working with a woman who lost her sight when she was twenty-seven, Beauchamp and his colleagues changed things up. Instead of firing all the electrodes necessary to represent a letter at once, they tried setting them off one after another. The woman was able to reproduce the patterns on a touch screen. Since then Dr. Beauchamp has extended this work to several other volunteers. He hopes by dramatically increasing the number of electrodes, this kind of "phosphene" vision will become much more detailed.

These are only individual letters, of course, hardly adequate for reading, but Dr. Beauchamp argues that these experiments open the door to creating "outlines of other common objects, such as faces, houses, or cars." We might even see a set of electronic emojis created specifically for people who have gone blind. (Unfortunately, this system appears not to work for people born without sight.)

We've come a long way from those ice age cave paintings to providing hope of restored vision to the blind—and phosphenes are the common thread.

Can you really learn to speed-read?

BEFORE WE TRY TO ANSWER THIS QUESTION, find the stopwatch on your smartphone, go back to the previous page, and time exactly how long it takes you to read. Go!

Now that you've timed yourself, let's do the math. There are about 200 words on that page, and reading speed is measured in words per minute (wpm). So if you took 40 seconds to finish that page, you were reading at a speed of about 300 wpm (200 x 60 ÷ 40). The average adult reads 200 to 300 wpm, so that's in the ballpark. If you took only 30 seconds, your reading speed was an impressive 400 wpm (200 x 60 ÷ 30).

Read-y, set, speed!

What's the rate of someone who claims to be a speed-reader? The numbers vary depending on who's doing the measuring, but most researchers agree it's somewhere between 3,000 and 10,000 wpm. So a true speed-reader would have steamed through that page in a maximum of 8 seconds and a minimum of just under 3! That is astounding and, yes, maybe something we'd all like to be able to do.

So that's the "speed" part. But what about the "reading"? Well, that's another story. There is much evidence showing that people who claim to be speed-readers are simply taking in very little of the information on the page. In fact, what they do should properly be described as "skimming," not "reading."

For decades, researchers have been trying to analyze exactly what speed-reading is. George Spache of the University of Florida was one of those researchers, and in 1962, he took on the challenge of evaluating an extraordinary claim. Evelyn Nielsen Wood of Evelyn Wood Reading Dynamics Institute in Washington, DC, said she could teach anyone to increase their reading speed by a factor of ten. There were reports of exceptional readers taking in anywhere from 8,000 to 12,000 wpm, or "reading" a whole page at a single glance.

How is that possible? One technique was described in the *New York Times*: "You turn the pages with your left hand and run your right hand down them with a gently caressing motion, letting your eye follow the right hand from top to bottom instead of reading across the lines from left to right." So the idea was *not* to read the way we're all taught, from left to right (for some languages), but rather to go straight down the page.

Spache wondered if it was even humanly possible to read that fast. He concluded it wasn't— because it's all a matter of time. Even when using Wood's technique, your eye has to stop and fix on a group of words, then jump to the next group and so on. Spache concluded that the human eye can take in only about three words at a time: one inside the so-called target area (which is about the width of your thumb when held at arm's length), and maybe one or two more just outside (if you're reading in English, that would be just to the right). He calculated that the maximum reading speed would be 900 words a minute, and that anything more than that involved omitting entire lines, and therefore can't be called reading. It's skimming.

Spache tested graduates of the Reading Dynamics Institute for speed and comprehension while tracking their eye movements at the same time. He found that for most, their reading speeds even after training were just slightly above normal, around 400 to 600 wpm. Those who skimmed text were much faster, more than 1,000 wpm, but they generally comprehended only about half the material on the page. The tracking showed that those who skimmed did indeed run their eyes down the page in a straight line, à la the *New York Times* article, meaning they took in only a tiny cluster of words in the middle of each line.

Not all researchers agreed with Spache's contention that speed-reading was mostly an illusion. Some argued that while a normal reader is trying to make sense of one sentence at a time, a speed-reader, who's reading down the middle of one page and up the next, is trying to combine scattered information from several sentences. That idea is that the reader can fill in the blanks from the context of what's being read. As Wood herself said, "Which would you rather do: eat a dish of rice kernel by kernel or take a spoonful to get a good taste?"

It sounds like a good idea, but it doesn't stand up. In one memorable 1983 study, two men who claimed to be able to read at an earth-shattering 100,000 wpm were given a college textbook and read it at a leisurely (for them) 15,000 to 30,000 wpm. Then they were tested on the material. They both failed. The author of the study concluded that their only real talent was their "extraordinary rate of page turning."

Did You Know . . . Woody Allen once said, "I took a course in speed-reading, learning to read straight down the middle of the page, and I was able to go through *War and Peace* in twenty minutes. It's about Russia."

More recently, a new technique called RSVP (rapid serial visual presentation) has emerged. It is supposed to enable great increases in reading speed. It works by presenting a single word at a time in rapid succession rather than a block of text. This is supposed to help your speed by eliminating the need to shift your eyes, reducing the number of unnecessary and even random moves. This method also claims to stop you from using your inner voice to read the words silently to yourself. (Although I tried it and it seemed to me that's *all* I did.) If you're curious, find one of the RSVP apps online and try it for yourself. But note that reading experts think that not being able to backtrack to a previous word or look ahead to the next makes this kind of reading more difficult, not easier.

This idea that reading is simply a repeating pattern involving a jump, a pause to take in words, another jump, and so on sells short the complexity of the task. To comprehend what you're reading, sometimes you need to read the same sentence more than once; it is an intricate piece of brainwork, a combination of rapid-fire eye movements and constant understanding. Doing it at superspeed? That's just a dream.

Why is the car radio so distracting?

IMAGINE DRIVING IN AN UNFAMILIAR NEIGHBORHOOD, looking for the address of a house you've never been to before. It's evening, so the lighting isn't the best. Half the houses have numbers you can't see or are missing them altogether. Put all these ingredients together and your brain is facing a challenge.

So you turn off the radio.

This isn't as bizarre as it sounds. There's a lot going on here that your brain must track:

remembering the address, trying to make out the house numbers, mentally comparing each number to the one you're looking for, *and* driving a car. Your brain is taxed by this deceptively simple undertaking.

This music is getting on my nerves!

The problem is that there's a limit to the amount of information your brain can deal with at any moment. In this case, you have to maintain the address in your short-term, or working, memory, even if you have it written on a slip of paper or recorded in a smartphone. From time to time, you'll have to take your eyes off the buildings to refresh that memory. Working memory doesn't last long—usually just enough to keep a phone number in your mind while you walk across the room to make a call. But keeping something in your mind does strain your mental resources.

You're also behind the wheel of a car. Thankfully, some of the effort of driving is managed by your unconscious. The human brain operates in both unconscious and conscious modes. Your conscious brain is responsible for every thought that's running through your head—any pieces of music, any memories, anything you think about for even a moment. Yet consciousness isn't what your brain does most. Of all the sensory information that flows into your head—in other words, everything you see, hear, touch, taste, and smell—only about one-millionth actually makes it into your consciousness. The stream of consciousness is really just a trickle.

The rest of that information ends up in your unconscious or is discarded. The unconscious mind takes care of everything you don't have to actively think about, from recognizing someone to signing your name. When you first learn to drive a car, you are using your conscious mind, thinking about everything, but that changes with experience. In our neighborhood drama, a significant percentage of the driving is handled automatically—unconsciously—because the mechanics of driving have been so well learned you don't actively think about them.

But that doesn't mean you don't have to pay some attention. You have to devote at least a bit of brainpower to avoiding a collision, whether it's with an unexpected pedestrian or the side mirror of a parked car. Even if you're driving very slowly, there's a risk of an accident because you're busy scanning the houses on one side of the street or the other. There's truth to the old adage that you steer where you look, so you have to be alert.

Science Fact! *In the 1880s, a French psychologist reported being able to recite one poem while writing out another at the same time, and to answer questions about both. Forty years ago, two Cornell University students decided to perform their own version of this experiment. They spent four months practicing reading a short story while writing out words that were dictated to them. The short stories were up to five thousand words long, and the dictated words were presented every six seconds. The students proved to be amazingly good at both understanding the stories and remembering the dictated words. The psychologists who ran the experiment remarked that it was hard to see where the upper limits for attention are for any of us.*

You're also visually challenged as you try to pick out, in low light, the address that you're juggling in your working memory. At night, even the clearest house number likely requires a second glance before you are sure it's the right one. If you could see the pupils of your eyes, you'd notice they're dilated—not just because of the darkness but also because of the mental effort you're exerting.

And then to all this you add the music or chatter from the radio. Your brain is already being bombarded with information—more information than you can readily take in. And while you're not really paying attention to the radio (you probably wouldn't be able to remember what was said or played), the sound is nonetheless a distraction. It demands attention that you are hard-pressed to give.

Psychologists haven't settled on one theory to explain situations like this. Irrelevant information must be screened out to prevent it from taking up too much of your attention. The question is, When? Does this happen as soon as the information is taken in, before any attention is wasted on it? Or not until the store of attention is almost drained? In this case, whatever is on the radio is clearly irrelevant, but at what point did it become a distraction? There's no way of knowing for sure, but it's pretty cool to think that you killed the radio just as you hit your attentional limit.

File too big.

Did You Know . . . When radios were first being put into cars in the 1930s, several American states toyed with the idea of barring drivers from listening while they were in motion. Connecticut even suggested a fine of $50, which would be equivalent to more than $800 today. But no laws were ever enacted. And one recent study showed that listening to music while driving had no significant effect on a driver's attention. Although if the driving is challenging, you might not remember the music!

Is the human body temperature dropping?

PEOPLE DON'T TRACK MUCH DATA about their bodies, other than height, weight, and temperature. Most of us know that normal body temperature is 37°C (98.6°F). It can be a touch lower or higher, but the safe range is pretty narrow. A fever of more than 40°C (104°F) requires medical attention, while a temperature below 35°C (95°F) is classed as hypothermia and is dangerous.

So that's why it was a surprise to many when a massive study of human body temperature from the mid-1800s to the present showed that on average, it has dropped by 0.6°C (1.08°F). That number had always seemed to be one thing that would never change, and yet we find out it has been changing steadily over a century and a half. The question is, Why?

Did You Know . . . During a July 1980 heat wave in Atlanta, a man named Willie Jones reached a temperature of 46°C (115°F) — and survived. He holds the Guinness World Record for highest recorded body temperature. Swedish girl Stella Berndtsson holds the record for lowest body temperature: 13°C (55°F). She was only seven at the time and had fallen off a steep cliff into the frigid water below. Today she is completely recovered.

The study was based on three major sets of data. The first covered Union soldiers from the Civil War and consisted of 83,900 measurements taken between 1860 and 1940, but of men only. The two remaining sets of data included both men and women. The second was of more than 15,000 people whose temperatures were measured between 1971 and 1975 as part of a national health survey in the United States. And the third was a Stanford University study that

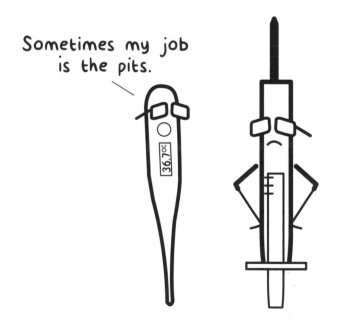

Sometimes my job is the pits.

ran between 2007 and 2017 and recorded 578,000 temperatures. Altogether, these data sets represented more than 677,000 times that someone said, "I'm going to take your temperature."

The sheer amount of data allowed the researchers to attach great precision to their analysis. So not only did they uncover the 0.6°C (1.08°F) overall drop, but they could track that decline in much finer time intervals. They learned, for instance, that in both men and women, body temperatures have been dropping by 0.03°C (0.05°F) per decade.

Still, the study's authors were concerned that different ways of taking temperature over that long span from 1860 to 2017—from a thermometer stuck in someone's armpit to one placed under the tongue—might have influenced the data. If anything, though, more modern methods should have bumped temperatures up, not down, because armpit measurements generally result in a slightly lower reading than those taken by mouth. Also, the fact that temperatures continued to drop steadily over relatively short time frames, like decades, suggested that any change in technique or technology had no significant impact.

So what accounts for the change? It might be more than one factor. For instance, you might suspect that body mass could play a role. It's well-known that average weight has increased steadily in the United States and other Western countries since the mid-nineteenth century—the result of better nutrition (or too much nutrition). The size of the average human is relevant because body temperature correlates to size: the bigger the animal, the higher the temperature. Yet in this case, the bigger animal has a lower body temperature.

Something called the thermoneutral temperature is also relevant. Historically, the so-called thermoneutral point—which is the air temperature at which the human body doesn't need to sweat or shiver to cool or warm itself—was said to be 28°C (82.4°F). Maintaining body temperature within the desired range simply by heat loss from the skin consumes something like 65 percent of the energy we burn. If the air temperature moves out of the thermoneutral point, we expend more energy, which in turn raises our body temperature slightly.

Did You Know . . . Body temperature is maintained by the hypothalamus in the brain. As a person's temperature rises, the hypothalamus increases the amount of blood flowing through vessels near the surface of the skin so that heat will be released to the air. It also induces sweating—the evaporation of sweat increases the amount of cooling. If the hypothalamus senses that body temperature is too low, it initiates shivering.

One suggestion made by the research team is that dramatic changes in the heating and cooling of homes have taken place since the nineteenth century, and that means we need to expend less energy to maintain a steady body temperature. Because we're expending less energy, we're generating less heat, and that has contributed to our cooler bodies.

The most interesting idea, though, is that the slight drop in body temperature over the past century and a half is related to a reduction in inflammation. Inflammation is a typical response to infection, and the researchers point out that infections we rarely think about today in North America, including tuberculosis, malaria, and gum disease, were once rampant—especially among Civil War veterans. (The Union soldiers surveyed in this study were not even given toothbrushes by the army!)

The management of medical care during the Civil War was clearly extremely difficult—more than 400,000 soldiers died of infectious diseases, meaning that two soldiers died from disease for every one who died from his war wounds. But from that time until now, medicine has advanced in many ways to curb inflammation. The introduction of vaccines, antibiotics, and even over-the-counter anti-inflammatory drugs like aspirin has helped reduce inflammation rates—and thus body temperatures as well.

Will the rising temperatures of climate change reverse this trend? Because the downward creep of body temperature is very slow, it will be challenging to track it into the future, but this study suggests that unless we spend more and more time in climate-controlled surroundings, then, yes, we might see a reversal of this trend that has lasted for more than a century.

History Mystery

Was Robin Hood a real person?

Surprisingly, this isn't a straightforward question with a simple yes for an answer. But, surely, you're probably thinking, the God-fearing gentleman outlaw—with his Sherwood Forest buddies Little John, Friar Tuck, Alan-a-Dale, Will Scarlett, and, of course, Maid Marian—did actually harass the sheriff of Nottingham, defy the villainous Prince John, and steal from the rich to give to the poor. With such a classic tale of good versus evil, it would be awkward to think someone made the whole thing up.

Unfortunately, nothing about this question is crystal clear, starting with the fact that there is no unambiguous proof of the existence of Robin or any of his Merry Men. The only person we can be certain did exist is the evil Prince (later King) John. We know a lot about John, but not much about any of the others.

Robin Hood, Robyn Hood, Robot Hood, and Big Hood

It's a famous name, but it turns out there were all kinds of Robin Hoods. Allowing for spelling (and likely pronunciation) differences, we can find evidence of Robert Hood, Robyn Hode, Robin Hod, and Robertus Hode. There's also William Robehod, Gilbert Robehod, and John Rabunhod. There's even a woman named Katherine Robynhod.

Is one of these *our* Robin Hood? The name alone isn't enough to go on, but there are other kinds of evidence. Scholars have turned to a series of rhymes—or ballads—sung by minstrels. These earliest-known references to Robin Hood first made it into print in the 1400s, but they had been around long before that. For instance, there were references to them in a long poem called *Piers Plowman*, written by William Langland in 1381. Apparently, they were already well-known by then, suggesting they dated back to before 1350—maybe even to the early 1300s. That would put Robin in Sherwood Forest in the late 1200s.

 **Did You Know . . .** Robin Hood was said to be a fantastic archer, renowned for being able to split the wand, or the peg anchoring the target, with an arrow. But oddly enough, he and his Merry Men apparently used their longbows not in battle but for hunting and sport. Odd because the longbow was a fearsome weapon of war. In one oft-told story, an arrow shot from a longbow was said to have passed through a man's thigh, pierced his saddle, and killed his horse.

Intriguing clues continue to turn up. In 2009, Julian Luxford, a professor of art history at St Andrews University in Scotland, found a handwritten Latin note in a medieval manuscript. It read, "Around this time, according to popular opinion, a certain outlaw named Robin Hood, with his accomplices, infested Sherwood and other law-abiding areas of England with continuous robberies." Luxford's convinced that the writer of the note meant for the phrase "around this time" to be clear from the events described elsewhere on the page. From that, Luxford reasons the time was the very late 1290s.

We have no idea how true the written versions of the rhymes are to the originals, and even the originals would constantly have been revised to suit the minstrels' audiences. But they do all tell the same familiar tale: Robin and his men repeatedly get the better of the sheriff, rob monks and other money-laden travelers, and give their ill-gotten gains away to those in need. But that can't be the whole story. For instance, Maid Marian does not appear in any of the rhymes—or in anything else written about Robin Hood for the next few hundred years.

Wait, where am I?

It's easy today to be misled into thinking that the many modern movies and TV shows about Robin are based on real events, but actually they represent centuries of tinkering with stories. It's probably impossible to trace the originals. It's amazing enough that the ballads have lasted at least seven hundred years—it would be too much to expect them not to have changed.

Did You Know . . . In the famous rhymes, Robin is unwell and visits his aunt, the prioress of Kirklees, to be bled. (Bloodletting was a standard medical practice at the time.) Unfortunately for Robin, his aunt had hatched a plot to murder him, and the bloodletting provided the perfect opportunity. The prioress simply didn't stop the bleeding (one of the more gruesome versions of the story said she used a bowl with a hole in it), and Robin died soon after.

Despite this overload of conflicting and vague information, many tantalizing questions remain. Such as where is Robin buried? Legend has it that Robin died in Kirklees in West Yorkshire. But before he expired, he is said to have asked Little John if he would help him shoot his last arrow. His friend agreed, and Robin said, "Bury me where my arrow falls." A symbolic last gesture from a great archer (although this story, like so many, appeared much later than the others).

Could we find that grave? Maybe take DNA samples to narrow the search for the real Robin? Actually, there *is* a gravestone on the Kirklees Estate. But in 2015, an American TV program called *Expedition Unknown* examined the gravesite—using ground-penetrating radar to image below the surface—and found no evidence there was anything there at all. More important, the grave is six football fields away from where Robin would likely have been in his last days, and a dying man—even one as skilled an archer as Robin Hood— couldn't possibly have shot an arrow that far.

And that's where it stands today. You can find many experts who are certain of Robin Hood's identity. The problem is that all their Robins are different.

Part 3
Oddities and Eccentricities

What makes some clowns scary?

THIS IS A PARTICULARLY CHALLENGING QUESTION because the history of clowns reveals that scariness seems to have gone in and out of fashion over time. Clowns have not always fit the modern stereotype of a cheerful, red-nosed, floppy-shoed Ronald McDonald in thick makeup and an oversize suit. There was Emmett Kelly, whose sad but unthreatening "Weary Willie" clown starred in circuses in the 1930s and 1940s. More recently, circus clowns were famous for stunts like cramming eight or nine clowns into a Volkswagen Beetle. They were goofy, not sinister.

But go back another fifty years and you find Canio, the clown in the opera *Pagliacci* (Italian for "clowns"), who stabs his wife and her lover to death onstage. In the last few years, we've returned to that—the goofy clown of the early twentieth century has been shouldered aside by the scary, murderous clown. So now we have Stephen King's novel *It*, in which

They're not really my size.

Pennywise the Dancing Clown kidnaps and eats little children. Then there are the various versions of the Joker, who morphed from a prankster in the 1950s and 1960s into the homicidal, disturbed man of the modern era. John Wayne Gacy, an actual serial killer known as the Killer Clown, appeared at children's parties and community fundraisers dressed as Pogo.

Did You Know . . . A recent study in Sheffield, England, showed that children generally dislike clowns. As a result, hospital administrators discontinued their practice of putting clown posters in children's rooms. In the words of one member of the research team, children found clowns "unknowable."

We can probably all agree that some clowns are creepy. But what makes them creepy? American psychologist Francis McAndrew has conducted a survey of creepiness, and he says it arises when we're confronting ambiguity. "Creepiness is all about not being able to figure out whether there is a threat," he explains. In social interactions, we look for regularity and predictability. We expect others to behave in certain ways, and we read social cues to determine if people are trustworthy. Mimicry is a good example: a little mimicry (you cross your legs and I cross mine) is friendly, but too much is creepy.

McAndrew's survey of creepiness identified some important traits and behaviors that might make others wary. They include everything from long fingers, greasy and unkempt hair, and bulging eyes to unpredictable laughter, pale skin, and bags under the eyes. There were other creepy behaviors—like asking to take pictures or talking too much about sex—but the emphasis on personal appearance, especially facial features and expressions, stands out. A clown exhibits several of these off-putting traits.

Did You Know . . . As part of Francis McAndrew's survey, volunteers rated the creepiness quotient of twenty-one different professions. Clown was judged the creepiest of all, finishing ahead of taxidermist, sex shop owner, and funeral director. I was happy to see that writer was only eleventh. The least creepy profession? Meteorologist.

Clowns may hang out in what's called the uncanny valley. This term was originally used to describe the discomfort people feel with robots that look almost human, but not quite. As long as robots are either square-headed and vintage like Rosie from *The Jetsons* or cute like R2-D2—in other words, as long as they clearly aren't human—we're comfortable with that. But a robot that looks almost human is unsettling to us. They've been described as "too dead-eyed to be believable, but too realistic to be cute."

But I *feel* human.

Exactly what causes the discomfort of the uncanny valley isn't clear. Some experiments have suggested that we don't like it when human and robotic qualities mix—for example, it's unnerving to hear a human speaking with a robotic voice or to see a prosthetic hand that looks human but has a skin texture that is clearly not. Other research has highlighted the reaction to unusual features, like a nonhuman eye in a human face, or even an eye positioned oddly in the socket. Both disturbed viewers. But there have also been several experiments that asked people to judge a series of faces as they gradually morphed from nonhuman into human. These tests failed to pinpoint the sudden drop in likability that would identify the uncanny valley. So it's not possible at the moment to define exactly what's happening at the meeting point between human and nonhuman.

Did You Know . . . The popular movie *The Polar Express* apparently created an uncanny valley feeling for many viewers, possibly because its kind-of-human computer-generated characters exhibited real human movements through motion capture.

But there's no shortage of theories as to why the uncanny valley might exist. One is that as we evolved, faces that were strangely different might have signaled the presence of disease and so would have been disturbing, even frightening, to us. Also, humans are innately more comfortable with faces that display symmetry and are in proportion. And while it might seem strange to be arguing an evolutionary basis for a discomfort with robots or clowns, there is evidence that the uncanny valley is also experienced by other species.

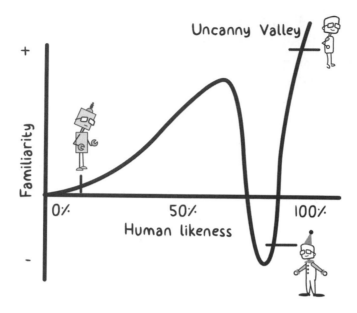

In one experiment, for example, macaque monkeys were presented with three versions of monkey faces: one natural and two computer generated. One of the computer-generated images was close to realistic, while the other was not. The monkeys expressed their preferences by spending much less time inspecting the nearly realistic face. The effect was enhanced when the faces were animated, lending support to the idea that natural versus unnatural movements are part of the uncanny valley. Apparently, it has ancient origins.

Clowns, though, may be the perfect embodiment of the uncanny valley. They are clearly human, but their facial features, especially the mouth and the eyes, are exaggerated to the point of being nonhuman, and their outsize movements might also contribute to their uncanniness. Whether they're cheery or threatening, the one consistent feature of clowns is that they stand apart from the rest of us: humans who aren't exactly human.

Do plants think and feel?

IF YOUR IDEA OF A PLANT is an inert green thing in a pot, you may think this question is absurd. But we now know there is much more to a plant's life than we thought. The question is, How far does that life go?

The claim that plants might have feelings or thoughts has had a rough ride. In the early 1970s, a book called *The Secret Life of Plants* created a sensation by claiming that plants react to threatening situations—such as being close to a human who has killed another plant—by ramping up the electrical activity in their leaves. The authors of the book were almost universally condemned by botanists for greatly exaggerating—and even inventing—the data they used to back up their claims. One irate scientist called the book a work of fiction.

Let's get to the root of this.

More recently, a German forester named Peter Wohlleben published *The Hidden Life of Trees*, a much more scientific book that looks at how plants communicate and what they may feel. Yet even Wohlleben irritated some plant scientists by describing trees as "anxious," having "positive feelings," and developing "friendships." Several other scientists researching plant communication also found themselves in hot water when they coined the term "plant neurobiology" for their work. They were rewarded with such vehement pushback that they changed the name of their proposed Society for Plant Neurobiology to the less offensive Society of Plant Signaling and Behavior.

Critics argue that since plants have nothing like a nervous system—no neurons, no synapses, and no neural circuits—it's impossible to study plant "neurobiology." But their opponents suggest that the fundamental divide between animals and plants might need reviewing. What's convincing some people that plants are closer to thinking beings than we've ever imagined? Let's start with the idea that they have feelings.

A study just published by a team of Israeli scientists showed that both tomato and tobacco plants emit brief ultrasonic noises when they're stressed, such as when they're cut or are drying out. The frequencies of the sounds were too high for human ears, but dogs could hear them, and they were loud enough to be heard from several meters away. Undamaged plants emitted no more than one sound every hour, but once damaged, both tomato and tobacco plants averaged no fewer than ten and as many as thirty-five per hour. Drying out triggered more sounds than being cut. Amazingly, a computerized system was able to tell the difference between sounds from a cut plant versus those from a dried one with 70 percent accuracy.

Even though this study has not yet been published in the scientific literature, it might revolutionize our thinking. Imagine walking through a forest and knowing that all the plants and trees around you are producing sounds in a kind of ultrasonic symphony.

This small study suggests there's a lot going on, but we don't hear it. But Dr. Suzanne Simard, a forest ecologist at the University of British Columbia, believes we don't *see* much of what's going on either. She has identified a vast underground network of fungi that she says trees in a forest use to connect with one another. She began her pioneering experiments years ago, when the idea of trees communicating with one another had not yet taken hold.

Stop all that barking.

Dr. Simard planted seedlings of both Douglas fir and paper birch trees in a forest, covered them with plastic bags, and injected the bags with two different varieties of carbon dioxide gas. Then she waited until the carbon dioxide was absorbed and had been processed into sugars by the trees. She found that the seedlings had exchanged the two varieties of carbon by releasing those sugars from their roots to the fungal network, where they traveled to the roots of neighboring plants. Two different species of trees were trading carbon.

But that's not all. Dr. Simard also discovered that when the Douglas firs were shaded and not photosynthesizing as efficiently, they received the bulk of the flow of sugars. But in the late fall, when the firs were still active but the birches had lost their leaves, the flow reversed. She even found that some trees, called mother trees, appeared to be taking preferential care of their offspring by transferring excess carbon specifically to them. Her work has demonstrated that the forest is an interconnected community, not a stand of individuals.

None of this so far proves that plants think or feel, but they certainly communicate, at least chemically and maybe acoustically, too. Still, it's tricky to interpret their actions. The mimosa plant, for example, is famous for folding its leaves when touched. This is apparently to protect the leaves from being damaged by whatever is doing the touching. What's interesting is that if a mimosa is repeatedly touched but never damaged, it will eventually stop responding. Even weeks later, a plant in the same setting will still not bother to fold its leaves if touched. Is it fair to say that it has learned it's not in danger?

Other plants are capable of releasing airborne chemicals to ward off attacks from leaf-eating caterpillars. The chemicals attract parasitic wasps, which lay their eggs on—or in—the caterpillars. When the eggs hatch, the wasp larvae consume the caterpillar. That ends the threat.

When the insect-eating Venus flytrap and other plants capable of movement are exposed to anesthetics, they are no longer able to respond to the presence of an insect. Does this mean they've been rendered unconscious by the anesthetic, as we are? Or does this simply demonstrate that the chemical effects of anesthetics include shutting down the signaling ability in the Venus flytrap?

Plants emit a variety of signals intended for other plants, and they are able to monitor changes in their environment. But do they make choices? Do they have feelings? Can they be frightened or in pain? If so, they haven't told us—yet. And while it is true that they don't have nervous systems like ours, it's undeniable that they have abilities we couldn't have imagined even a few decades ago.

What makes dogs so cute?

OUR WORLD HAS MORE DOMESTICATED ANIMALS than wild ones. But all those cats, dogs, and familiar barnyard animals—cows, sheep, pigs, and horses—had to come from somewhere. Each has a wild ancestor. For pigs, it's the boar. And for dogs, the wolf.

Dogs were domesticated from wolves perhaps 30,000 years ago. Domestication is what's made them different. Those ancestral wolves were similar but not identical to today's *Canis lupus*, or gray wolf. Comparing modern wolves to dogs might reveal what attributes made the wolves suitable for domestication in the first place, and what features combine to make them such good human companions. How much wolf is still there in the dog? This question is perfect for the dog because it can actually be tested.

It's all about eye contact.

For instance, a recent discovery in dog-versus-wolf research is a muscle that, so they say, makes dogs cute. You don't just stumble on things like this. At the University of Portsmouth's Dog Cognition Centre, a team of British researchers, led by biological scientist Juliane Kaminski, already knew a lot about dog-to-human communication. In particular, they knew that gaze is supreme. Dogs seek out visual contact with humans and respond to it. They are able to read the emotions and intentions of their human companions. This includes even simple things like understanding that a pointed finger is suggesting there's something of interest not at the fingertip but in that *direction*. Still, a dog will likely fail to follow that pointing finger unless it can see where the human is looking. Dogs learn to establish eye contact at an early age, and if they're unable to solve a problem on their own, their first reaction is to look at a human for clues.

This relationship has a chemical basis: when a dog and a human gaze into each other's eyes, it raises levels of the hormone oxytocin in both species, just as it does between mother and infant. Kaminski and her team also knew that both dogs and their owners prefer lingering looks over glancing eye contact, and that dogs will do a trick where they make their eyes seem bigger, cuter, more like a puppy's. Humans seem to fall for this: one study of dogs at a shelter showed that those who performed the eye trick were more likely to be adopted.

This is the answer to the original question: What makes dogs so cute? They're able to do this using a set of four muscles, two beside each eye. Each muscle is attached at one end to a ring of muscles that circle the eye, and at the other end to the brow above. When these four muscles contract, the dog's eyes open wider and take on a pleading "Look how cute I am" expression. (It's not me claiming they're cute—most descriptions of this research use the word.) These muscles allow a dog's eyes to change shape in a way that fits exactly with what humans like to see.

This is all part of a strategy to take advantage of the fact that humans prefer animals (and other humans!) that are more baby-like. In general, this means higher foreheads, bigger eyes, and shorter muzzles. This applies not just to dogs—we like the look in all our favorite animals. Mickey Mouse underwent cosmetic surgery from his first appearance in the 1928 short film *Steamboat Willie* to the Mickey we all know and love today. What were the changes? Bigger eyes, a shorter muzzle, and a higher forehead. Teddy bears have evolved in the same way.

Science _Fact!_ Starting in the 1950s, Soviet scientist Dimitri Belyaev conducted a series of experiments with silver foxes and demonstrated that many physical differences arise unexpectedly during domestication. Belyaev tested each generation of foxes for tameness. Would they retreat from people and try to bite, or approach and be friendly? Only the tame ones were allowed to breed. But even though tameness was the only feature being selected, the foxes changed in other ways as the generations passed. Their ears became floppy, their tails curly, and their fur changed color. They looked and acted different, wagging their tails and licking their trainers' hands. In other words, they became more like dogs.

What's striking is that modern wolves lack these eye muscles. They have some fibrous tissue with sparse muscle fibers in the same location, but they can't reshape their eyes the way dogs can. When wolves and dogs began to diverge those tens of thousands of years ago, it was before the existence of farming and the first cities. Long ago, but not a lot of time to turn connective tissue into muscle—unless there's a pronounced evolutionary advantage to doing so. Interestingly, of the breeds of dogs the researchers dissected to look for the muscle, only the Siberian husky, the most wolflike, lacked it.

Did You Know . . . While dogs are much better adapted to humans than wolves are, wolf pups do display some doggy behavior. For example, some will play fetch with a tennis ball—even if they've never seen the game before.

How did these eye muscles develop so rapidly, then? It's tempting to think that they were so crucial in establishing the social bond with humans that only the dogs who had the genes for them flourished, establishing forever the "cute eye" of the modern dog. Domestication is a powerful force and has prompted behaviors that allow dogs to connect to us socially. But this latest research indicates that even their anatomy has changed since they left the wild and took up with us.

Doggo   Pupperino   Floofer

Are UFOs a real thing?

YES, UNIDENTIFIED FLYING OBJECTS ARE A REAL THING—if by that you mean they are a topic of conversation, and people claim to have seen them. But if you mean they're something that really exists but cannot be identified, the answer is . . . possibly. And if you venture even further and define a UFO as an alien spacecraft, then the answer has to be no. At least, not as far as anyone can tell.

For millennia, people have looked up and seen strange things in the sky, but the modern interest in UFOs took flight in the late 1940s and early 1950s. Some sightings were hoaxes, but the vast majority could be explained as either high-altitude clouds, rogue weather balloons, the planets Jupiter or Venus, hurtling meteors, or the products of overactive imaginations. In Roswell, New Mexico, the crash of a weather balloon attracted believers who suspected a downed UFO. And Area 51, a secretive American military testing site in Nevada, became the darling of conspiracy theorists when parts of the rumored alien spacecraft from Roswell were supposedly taken there.

But some strange sightings do remain unexplained to this day. For example, on August 25, 1951, in Lubbock, Texas, several college professors were sitting outside enjoying a warm summer evening when suddenly a V-shaped formation of lights passed directly overhead. Over the following days, the lights appeared again and again, not just to the professors but to others, including a young man named Carl Hart Jr., who photographed them. At one point, the lights passed over low-level clouds, allowing witnesses to estimate their speed at something close to 1,000 kilometers (600 miles) per hour.

The US government sent an air force captain, Edward J. Ruppelt, to investigate. He was the lead researcher for Project Blue Book, a program that looked into UFO reports. Ruppelt considered several explanations for what came to be known as the Lubbock Lights, ranging from flocks of birds illuminated by streetlights—unlikely given that no birds can fly anywhere near the reported speed—to a type of tailless aircraft called the flying wing.

But in the end—and this is the strangest part of the story—Ruppelt claimed that the lights had been "positively identified as a very commonplace and easily explainable natural phenomenon." And that phenomenon was . . . what? Astonishingly, he went on to say that he couldn't divulge anything more because the scientist who had solved the mystery wanted to remain anonymous. Ruppelt died in 1960, and no one else has ever been able to explain what the Lubbock Lights might have been.

UFOs aren't limited to the United States. Canada's most famous case is the Falcon Lake Incident. On May 20, 1967, amateur geologist Stefan Michalak was prospecting for quartz in the rocks in eastern Manitoba when he saw two mystery objects in the sky, one of which landed on the ground nearby. Thinking it was an American military aircraft, Michalak sat down and sketched it. After about half an hour, he decided to approach: The air around the saucer-shaped object smelled of sulfur, and it was hot to the touch. Michalak could hear a whirring sound and was burned by hot gas emitted from tiny holes in the surface. Before the object took off, he was even able to see through a doorway into the interior.

Michalak was treated in hospital for first-degree burns on his stomach, and he also had a psychological evaluation at the Mayo Clinic, confirming there was no evidence he was either lying or deluded. (The doctor jokingly concluded that Michalak was "very down-to-earth.") No one has ever been able to explain what he saw or what happened, but later a piece of radioactive metal was found at the site.

For years, the UFO scene seemed relatively quiet, but just recently, a sensational story surfaced about American navy pilots who spotted a weird object off the coast of Southern California back in 2004. That object has come to be known as the Tic Tac because its shape resembled that of those oblong candies. Pilot Dave Fravor saw it first, and he said it was unusual because it was hovering with no apparent evidence of wings or engine or exhaust. About an hour later, a second pilot, Chad Underwood, picked it up on radar when it was still too far away for him to actually lay eyes on it. He was able to lock onto it with an infrared camera and record a little more than a minute's worth of video.

That video shows an object, seemingly about 12 meters (40 feet) long, hovering more or less steadily in midair. Near the end of the clip, it suddenly darts off to the left and disappears. Fravor said it was "like nothing I've ever seen," and Underwood explained that it was "behaving in ways that aren't physically normal." He said, "That's what caught my eye. . . . It was going from, like, fifty thousand feet to, you know, a hundred feet in like seconds, which is not possible."

What are we to make of this—or any other UFO sighting? The Tic Tac is probably the most interesting because there's video evidence and the witnesses were highly trained, careful observers. So what could the Tic Tac be? Could it have been a glitch in the infrared camera? If so, why did Fravor say he actually saw the object? Could it have been some secret American military aircraft? But these were *navy* pilots, so wouldn't they have known what planes the military had? Also, in its rapid acceleration and tremendously high flight ceiling, the Tic Tac would have been more advanced than any other conceivable US aircraft.

In early 2020, the US Navy announced that it possessed top-secret information about UFOs that it wouldn't release because it could cause "exceptionally grave damage to the National Security of the United States." That may speak more to the technology used to detect these objects than to the nature of the objects themselves. Then a couple of months later, the US Department of Defense officially released these videos (which had previously been leaked), apparently to prove they weren't keeping any part of them secret. It's all tantalizing.

Yes, there's a chance that intelligent civilizations are out there in the galaxy—just in the last twenty years, we've learned that thousands of previously unknown planets are orbiting distant stars. If such aliens exist and are more technologically advanced than we are, a Tic Tac–type aircraft could be real. But why would they come here? What were they doing? And why haven't we seen them again? You'll have to come up with your own answers—at least until the next sighting.

Is "huh" a word?

WHEN YOU THINK OF HOW OFTEN you hear someone say "Huh?" and in how many contexts, it's clear that whatever it is, "huh" has an important conversational role to play. According to linguists, its most common use is as what's called an "other-initiated repair," a way of fixing something that's been missed in a conversation.

Picture yourself listening to someone. If you miss something that person just said, and you need to know what it was—and quickly—you'll blurt out, "Huh?" The speaker will then respond by repeating or clarifying what was said, and the conversation will sail on. That's an other-initiated repair.

Word to the wise...

But does that make "huh" a word, or is it just a sound? To qualify as a word a sound has to be produced deliberately as part of a language. A sneeze doesn't count as a word. Nor does that grunt you make when hitting a tennis ball. Neither of those sounds is heard regularly as part of a conversation. But "huh" is—as we just saw.

Researchers in the Netherlands examined thirty-one languages from around the world and found that each one has a word similar to "huh." They were careful to avoid languages that are closely related, like English, French, Italian, and Spanish. Each of those descends from the same ancient language. It wouldn't be a surprise to discover that languages descended from a common ancestor had all "inherited" a version of "huh." Instead, the researchers looked at languages as distinct as Murrinh-patha (an Aboriginal language from Australia), Tzotzil (a Mayan language from Mexico), and Mandarin Chinese.

This study showed that "huh" has almost the same sound and the same meaning no matter which language it comes from. This is unusual because words with the same meaning in different languages often sound completely unalike. The English word "dog," for example, is *inu* in Japanese, *chien* in French, *sobaka* in Russian, and *yawi* in Duna (a language spoken in Papua New Guinea). "Huh," on the other hand, is consistent around the globe. Perhaps the question of this chapter should really have been "Is 'huh' a *universal* word?"

dog, inu, chien, sobaka, and yawi

Did You Know . . . As an other-initiated repair, the word "huh" stands on its own with a question mark. But it's also an exclamation. You'll also hear it used without a question mark to express amazement or surprise: "Did you know that she walked ten thousand steps a day for three months?" "Huh! " Or it can be offered as a series, with rising loudness and pitch, to express anger or issue a challenge: "Oh, you will, will you?" "Yeah, I will." "And then what, huh? Huh? HUH?" Finally, it can be used at the end of a sentence to invite agreement: "It's a scorcher today, huh?" The Dutch researchers concluded that all these uses were related to the primary one, but it's too soon to say exactly how.

What's unusual about the findings of the Dutch researchers is that to qualify as a word, an expression has to fit into the sounds of its language. If every "huh," no matter where it's spoken, is roughly the same, then it is more like a sound than a word. But when the Dutch linguists probed the acoustic details of every "huh" they examined, they found that each differed subtly to fit the general soundscape of its language. "Huh" also fits with the development of language: infants don't say "huh" and toddlers say it only sometimes, but once language is fully established, "huh" makes regular appearances.

"Huh" is short and sharp, suggesting that its form fits perfectly with its purpose. Conversations are fast; if you miss something, it's gone for good. But if you blurt a "huh," the speaker will immediately go back and repeat the piece you need—and "huh" has done its work. It's also easy to say quickly. In languages around the world, "huh" can be voiced when the tongue is relaxed in the mouth, making it easy to articulate whenever the conversational need arises. That minimizes the time that would be lost in planning and executing a more complicated sound.

Even so, there's always a slight pause before "huh" is uttered in a conversation—about eight-tenths of a second (versus two-tenths for normal conversational turn-taking). The listener needs that extra time to realize that something has been missed.

Did You Know . . . The word "eh" has not been analyzed to the same depth, so we can't yet claim that it's universal. But like "huh," it's an interjection that can mean different things. It can be used to ask questions, express surprise, or encourage agreement: "That's the way it goes, eh?" And among Canadians, it sometimes seems to appear at the end of almost every sentence, often meaning not much at all.

huh?

Out for a walk, eh?

Thanks to these Dutch researchers, then, we know that "huh" sounds and means the same in many, many languages that do not share a common ancestor, meaning that somehow they have arrived at this conversational solution—this other-initiated repair—independently. In biology, when two unrelated species hit on the same solution to an environmental challenge, it's called convergent evolution. The North American wolf or coyote, and the extinct marsupial, the Tasmanian tiger of Australia, look much the same, not because they're closely related—they're not—but because their way of life is similar. (It was called a Tasmanian tiger because it had stripes, but in its habits and behavior, it was much more like a wolf.) You'd have to go back 160 million years to find a common ancestor for the two, but both species chased similar prey and lived in similar environments. They faced the same challenges and evolved the same way to respond to those challenges.

According to the Dutch linguists, the same is true for "huh." Its primary purpose—in all the languages studied—is to repair breaks in conversation; it serves that purpose and does so quickly and easily. Every language needs these conversational Band-Aids, a sound that pauses the exchange when something is missed and allows the lost piece to be found.

Is there a mystery planet out there?

Just to be clear, this question is not about planets around other stars in the Milky Way galaxy—there are billions of those. This is about whether we have a ninth planet in our solar system, somewhere way out there beyond all the others.

It'd be nice to have another planet. After all, in 2006 Pluto was reclassified as a dwarf planet, a celestial body that is one rung below full planetary status. The definition of a dwarf planet is quite technical, but one of the key criteria is that it lacks the gravitational power (because it's small) to sweep up all the celestial objects and debris in its immediate neighborhood. Pluto is surrounded by that kind of stuff, so it was demoted to dwarf status.

1, 2, 3...

But now there's hope for a replacement! For the last several years, astronomers have been gathering data that suggest there's another ninth planet out there. And when I say "out there," I mean way, way out there—far beyond the eighth planet, Neptune.

Searching for Planet Nine, as it's called, involves a lot more than just booking telescope time and scanning the heavens. We have sent so many space probes and telescopes into the solar system that our extended neighborhood—everything from the sun out to Neptune—is pretty thoroughly studied and mapped. Any new planet will be much farther out than that, and therefore not even remotely visible to the naked eye and a minuscule target even if you have an extremely powerful telescope. It's possible to find it, but only if you have a very good idea of where to look.

So if it's not readily visible, why do astronomers think Planet Nine exists? Again, because of gravity. When you see models of the solar system, everything looks neat and orderly, with the planets moving smoothly along elliptical (not circular) orbits around the sun, like clock hands around a dial. But that's deceiving. Not only does the sun hold all the planets in its gravitational thrall, but the planets influence each other, too. For instance, both Jupiter and Venus affect Earth's orbit and contribute to changes in its climate over hundreds of thousands of years. If there's a planet on the outskirts of the solar system, it, too, might affect other nearby space objects. And such orbital anomalies have indeed been seen.

To understand what these anomalies are and why they matter, you first have to know that there's much more to the solar system than the eight planets, Pluto, and a handful of other dwarf planets. For instance, there's the asteroid belt between Mars and Jupiter, where scores of irregularly shaped rocky objects orbit, some of them occasionally hurtling relatively close to Earth. And out beyond the edge of the solar system, past Neptune's orbit, is the Kuiper Belt, a collection of hundreds of millions of icy objects.

Most of the millions of objects out there dutifully orbit the sun in roughly the same plane as the planets, tracing paths like the grooves on a vinyl record. But there is a cluster of objects whose orbits are steeply inclined to that plane, and such independence is hard to explain unless there's a massive celestial body whose gravity has driven them out of their normal routine. The wonderful thing about this is that the mathematics of gravity predicted such a set of objects would be there if indeed there was a massive Planet Nine—and there they were!

Once these outliers were observed, it became possible using astronomical calculations to make some educated guesses about the nature of the still-hypothetical Planet Nine. Currently, astronomers believe that it is an icy giant planet, as much as ten times the mass of Earth, with an enormous orbit. Planet Nine's year—the time it takes to complete one orbit around the sun—might be equivalent to ten or even twenty thousand Earth years.

A planet ten times the mass of the Earth isn't that remarkable—Jupiter is more than three hundred times the mass of Earth, and Saturn is ninety-five times its mass. Even Uranus and Neptune are close to fifteen times more massive. Venus, Mercury, and Mars are all smaller. Planet Nine would be a nice middle-of-the-road addition to our collection.

If it's out there, how will we ever find it? Astronomers have been searching places where they think it might be using the Japanese Subaru Telescope in Hawaii. It's powerful and can also scan large areas of sky. If Subaru doesn't find Planet Nine, then the search will switch over to the Large Synoptic Survey Telescope, which will come online in Chile in the next few years. And if we do find it, how do we know it will qualify as a planet? Because if its gravity is strong enough to tilt the orbits of those far-off space objects, it will surely have cleared the area immediately around it.

Finding Planet Nine would be exciting, but it would play havoc with scale models of the solar system. One of my favorites begins with the sun as a bowling ball. On that scale, Earth would be a peppercorn about 24 meters (79 feet) away. Neptune, represented by a coffee bean, would be more than seven football fields in the distance. But Planet Nine? Until we find it, we can't know what object would stand in for it (maybe a bead?), but we do know it would have to be at least 14 kilometers (almost 9 miles) away.

What is the deer-in-the-headlights phenomenon?

THE EXPRESSION "CAUGHT LIKE A DEER IN THE HEADLIGHTS" usually means an animal—or a human—has found itself in danger so extreme it freezes, unable to respond. This sometimes happens when drivers come upon a deer in the road at night, and instead of fleeing, the deer stands and stares at the oncoming car. There are no statistics kept for how often this happens, but in the United States alone, there are well over a million collisions between vehicles and deer every year. It's safe to assume that many of those were the result of a deer standing motionless on the road.

Why would a deer do this? One suggestion is that because they're generally active at dawn and at dusk, when light levels are low, their eyes have adapted to dim light, and the sudden glare of headlights temporarily blinds them. There is some

Oh, deer, I see the problem.

science to back this up. One study found that if the headlights on a truck were changed to produce light more closely in line with the deer's visual acuity, the animal would leave the road earlier. So there might be some truth to this, but it's definitely not the whole story.

Science _Fiction!_ Each year, deer are involved in more than 1.2 million vehicle collisions in North America, resulting in more than 200 human deaths and nearly 30,000 injuries and close to $2 billion in costs. One study suggested that if deer-hunting cougars were reintroduced to the eastern United States, they could reduce deer-car collisions by 22 percent and prevent 21,000 injuries and 155 deaths to humans over thirty years.

Another suggestion is that deer keep still as a defensive tactic. Again, there is some science here. We know that many animals freeze at the sight of a predator, prompting some scientists to argue that the fight-or-flight response should be renamed freeze, flight, or fight.

Freezing might seem like a risky choice, but it depends on the prey and the predator. Any animals that rely on camouflage—like fawns with their spotted coats—are usually better off freezing at the first sign of danger. By blending with the background, they may escape notice. But if a predator's vision or hearing is highly dependent on detecting movement, even an uncamouflaged animal would benefit by remaining still. A vole that freezes as an owl slowly cruises by might get away because the owl's vision is not acute enough to spot an unmoving lump in a field. And the owl is listening for sounds, and the vole isn't making any.

I've got cold feet.

Sometimes an animal will vary its response (freeze or flee) depending on the situation. The wood mouse (aka the field mouse) sometimes freezes—or when confronted with the weasel-like stoat, it may leap instead—but it usually just runs from predators. Freezing is a defensive behavior, carefully calibrated to the hunting style of the predator. Sometimes it works and sometimes it doesn't, but it must work more often than not, given how prevalent it is in the animal world.

Did You Know . . . Age can make a difference. Young white-tailed deer are prone to freeze when they're in danger, while adults are more likely to flee.

Even fruit flies act like deer in headlights, but in their case, we know why: the neurotransmitter molecule serotonin is responsible. When a fruit fly is startled by a sudden vibration, serotonin is released in the fly's spinal cord, and the insect comes to a dead stop. Researchers speculate that frozen moment gives the fly time to assess the unexpected situation and decide how to respond.

Do humans act like deer in headlights? We do have serotonin in our brains, but that doesn't mean it plays the same role as it does in fruit flies. Primates like macaque monkeys, which are closely related to humans, freeze when startled, too, but do we?

Several experiments have tried to measure whether people's bodies stiffen when they see unpleasant, scary, or disgusting pictures. This isn't easy to do: even if we think we're standing perfectly still, there's always a small amount of swaying or twitching going on, and our muscles are constantly tensing and relaxing to counteract that movement. In some studies, subjects were asked to stand on a stabilometric platform—a device that tracks the bewildering number of these tiny movements. They were then shown pictures of sports, everyday objects, and horrific injuries and death.

Researchers noted that their normal sway declined abruptly when the subjects looked at the unpleasant pictures, suggesting this was a subtle version of freezing. They weren't completely motionless, but they stilled themselves. At the same time, their heart rates slowed, and a drop in heart rate is common in animals who freeze as a defense. This experiment has been replicated using film footage and even a series of faces (happy, neutral, and angry), with similar results.

Psychologists are fascinated with this evidence that even humans can act like deer in the headlights in response to a life-threatening situation. But the reaction can vary significantly from person to person. One study involving firefighters demonstrated that when they were shown unpleasant pictures related to their work, they didn't freeze at all, while nonfirefighters did, suggesting that exposure reduces the impact of fear. I don't know whether this is connected to the idea that freezing buys time to prepare for action. But if it is, experienced firefighters don't need that time.

The idea that freezing might diminish with experience was also demonstrated in a study of war veterans with post-traumatic stress disorder. Despite the traumas they had endured, they did not freeze at all in response to unpleasant images.

But it's likely that most of the rest of us would freeze, however slightly, if shown disturbing pictures. We might not even be aware of it, but when it happens, we're all deer in the headlights—for better or worse.

What was the worst year ever?

IF YOU WERE A DINOSAUR, the year the asteroid hit would easily be the worst one ever. But for us humans, the year 536 CE has to be a contender. It has some serious competition, though—including 1918, when those who hadn't been killed in the First World War began dying in the influenza pandemic, and 1845, the start of the Irish Potato Famine. But 536 was even worse.

Oh, that's easy.

The aftereffects of a seriously bad year are felt through the following few years, so 536 holds the title not just because it was uniquely bad but because it continued to impact the years that followed. So, for instance, in 540, a second volcanic eruption shook the world, and in 541, the Justinian Plague hit, killing millions. Each of these three events led to immense loss of life and contributed to the collapse of economies and even to the destruction of empires.

First the volcanic eruption of 536. Contemporary writers in Europe and Asia had no idea there had even been one. How could they know? The best candidate for the volcano is Ilopango, in El Salvador, a part of the world unknown to Europeans and Asians at the time. All they knew was what hit them. A "dry" fog of ash and sulfate compounds from the eruption spread over Europe, blocking out the sun. One writer in the Middle East reported, "The sun was dark and its darkness lasted for eighteen months; each day it shone for about four hours and still this light was only a feeble shadow. . . . The fruits did not ripen." The sun was likely only one-tenth as bright as it is now, and an analysis of tree rings from the time shows that the average temperature dropped by as much as 2.5°C (4.5°F). We know this because when growing seasons are cold, tree growth slows and tree rings are narrower.

By 540, the fog had dissipated and temperatures began to recover, but then another eruption sent more ash drifting over Europe and Asia. Another dramatic fall in temperatures ensued. These two eruptions combined to make this entire decade the coldest in two thousand years. This wasn't just a series of cloudy days. It made life next to impossible for people whose life was hard enough already. The failure of crops and the bitterness of the cold contributed to widespread misery.

The prolonged "winter" and the years of deprivation and starvation that followed would be enough to qualify this short period as one of the most miserable on record, but the fates weren't done yet. In 541, the Plague of Justinian began to hit. This is the same kind of plague—identical bacterium and method of spread—as the infamous Black Death of the mid-1300s, so the Justinian version, named after the Roman emperor of the time, is considered the first pandemic plague in history.

Whether it ranked with the Black Death in its destructive impact isn't clear. Historians have traditionally believed that the Plague of Justinian killed tens of millions of people—perhaps as many as 50–100 million—beginning in 541 and continuing on and off for two centuries after that. But recently, scientists have cast doubt on those numbers, assembling evidence showing that in many ways life seemed to go on as normal throughout that time. There is little evidence that farming was abandoned; if that had happened, the varieties of pollen would have changed through the years (this was definitely an effect of the Black Death). Economies, as measured by the abundance of coins and the values of currencies, seemed fine. Mass burials didn't significantly increase.

So while the overall impact of the Justinian Plague is uncertain, it was still devastating that the eruptions of 536 and 540 immediately preceded the outbreak of plague. You can be sure there were many places where this would be the last straw. Such as in Antioch, a major city in what is now Turkey. The city endured the eruptions, then a "great" earthquake in 536 as well. Then in 540, the Sasanian emperor Khosrow captured the city. And there was the plague.

What this research reveals is just how precisely scientists can investigate even the distant past. The evidence for the dry fog, for example, was found by drilling ice cores in the Swiss Alps. An ice core is ringed from top to bottom, with each ring representing a season as the ice is laid down. It's a brilliant calendar, but it also records the chemistry of the atmosphere over time. When a volcano erupts, sulfur and other chemicals explode into the atmosphere, then slowly settle and can be found in the ice centuries later. In one ice core from the Alps, intervals as short as two weeks from centuries ago can be distinguished. So scientists can know what two weeks' worth of snowfall was like in 536!

But these cores are more than just a record of the weather. Laura Hartman, a graduate student at the University of Maine, found, in the core from the year 536, tiny particles of volcanic glass that resembled volcanic rock previously found in Iceland. That seemed to suggest that Iceland could have been the source of the first eruption, too, although some scientists suspect it was a volcano in central America instead.

Another team, using tree rings rather than ice cores, established that the period following the eruption of 536 was so cold that it ranks with the famous Little Ice Age of the 1600s. The scientists examined two millennia's worth of tree rings and found what they called a period of "unprecedented long-lasting and spatially synchronized cooling." They were bold enough to suggest that the fallout from these eruptions fostered the dispersal of Slavic people and political unrest in China.

It's a demonstration of the amazing reach of science that ice cores recovered today can illuminate life in antiquity.

What are the odds your lost wallet will be returned?

A FABULOUS INTERNATIONAL EXPERIMENT took on this exact question, and you will be impressed to see how far the researchers went to ensure that they had reliable results.

This experiment explored what happens when altruism comes up against self-interest. In it, thirteen research assistants fanned out across 355 cities in 40 countries, armed with a stunning 17,030 "lost" wallets. Each wallet was made of transparent plastic and contained three business cards, a grocery list, and a key. Some also included a small amount of money—the equivalent of about $15 in the local currency.

Island of Lost Wallets

In each city, the assistants "found" as many as fifty wallets and turned them in to banks, post offices, hotels, theaters, museums, and police stations. (They chose these because they can be found in every country and usually have a public reception area.) A researcher would walk in with a wallet, push it toward the receptionist, and say, "I found this on the street around the corner. Somebody must have lost it. I'm in a hurry and have to go. Can you please take care of it?" The assistant would then walk out, having left no contact information, and the experimenters would wait to see what happened next.

Because the wallets were made of clear plastic, the recipients could easily see what was inside. And the business cards included contact information, like an email address, giving them an opportunity to get in touch with the rightful owners. Every attempt at contact made in the first one hundred days was tabulated.

Much to their surprise, the researchers found that people were *more* likely to return a wallet with money than one without—by an average of 51 percent to 40. This result was seen clearly in 38 of the 40 countries tested.

The experimenters wondered if the results simply reflected that there wasn't enough money in the wallets, that the stakes weren't high enough. So they upped the ante. They sent research assistants in the United States, Poland, and the United Kingdom back out with a third wallet— the "big money" wallet—loaded with seven times as much cash as the original, or somewhere between $90 and $100. This time, people returned the "no money" wallet 46 percent of the time, the "small money" wallet 61 percent of the time, and the "big money" wallet a jaw-dropping 72 percent of the time.

What would you have expected? Most of us would probably have guessed the opposite: that more people would hang on to a wallet with money (especially "big money") and return one with none. In fact, the research team polled both random citizens and professional accountants and asked them what they expected would happen. Most thought the wallet with no money would be returned more often.

This was a powerful study, but there are always shortcomings. I can't help wondering if the setup might have skewed the results: the people who received the wallets held down jobs at respectable institutions, and my hunch is that people in those settings might be more likely to behave honestly. Results from handing a wallet to a pickpocket on the street would probably be different.

Regardless, it's rare that we have our expectations so thoroughly contradicted. So what's going on here? Was this a simple display of altruism? Perhaps. Researchers did find that the presence of a key was crucial. While a grocery list and business cards can easily be replaced, a lost key is an annoying hassle. Maybe the finders were sensitive to that.

Lock, I can get us out of this.

Did You Know . . . Wallets without keys were used in another small-scale experiment in the same three countries as the "big money" wallet. Researchers in that study found that a wallet with a key was 10 percent more likely to be returned than one without.

Altruism does require thinking about returning the wallet and actually doing it, and some people can't be bothered, but obviously a significant number, somewhere around 40 percent, take action. Altruism was crucial, but ultimately, the experimenters concluded that the recipients of the wallets were motivated by a combination of that and guilt—they didn't want to see themselves as thieves. The study asked whether keeping a lost wallet would feel like stealing, and those surveyed consistently said yes. Interestingly, though, most said that keeping a "small money" wallet felt more like stealing than keeping a wallet with no cash, and keeping the "big money" wallet was even worse. Based on the final results, we can conclude that close to three-quarters of wallet finders don't want to feel like thieves.

One thing that might interfere with acting on one's altruism is, of course, self-interest. It's easy to imagine someone looking at a wallet with $100 in it and keeping it. That happens, but not nearly as much as you'd think. If you're wondering whether some people returned the wallets but took the money out first, the answer is yes. That did happen, but only about 2 percent of the time.

I've had my own experience with this. I once lost my wallet in a hotel parking lot, and two days later, I got a call to say it had been turned in. The wallet had had some cash in it—more than described in these experiments. When I opened it, I discovered that someone had removed 96 percent of the money! So the finder was altruistic in returning the wallet—which had my driver's license, health card, and several other items important to me but useless to anyone else—but he or she was self-interested enough to take almost all the money. In this case at least, the fear of feeling like a thief was apparently overwhelmed by the temptation of a payday.

Are banana peels really slippery?

SLIPPING ON A BANANA PEEL was a classic gag of early-twentieth-century slapstick comedy. Movie stars like Buster Keaton and Charlie Chaplin and burlesque comedians like "Sliding" Billy Watson would slip, slide, and skid their way across screen and stage.

The banana peel gag was somewhat based on real life. By the late 1800s, bananas had become an extremely popular fruit, and because garbage disposal and urban composting weren't what they are today, sidewalks were littered with rotting peels, making them an actual hazard. But why were banana peels more dangerous than any other garbage you might encounter on a sidewalk? Are they really that slippery?

Some find me appeeling.

There's no better place to search for an answer to this question than a tribology journal. (Tribology is the science of interacting surfaces in relative motion.) And sure enough, in the journal *Tribology Online*, a team of Japanese tribologists published a study called "Frictional Coefficient Under Banana Skin." Recognizing the banana peel's reputation as one of the slipperiest materials known to man, the Japanese researchers set out to measure just how slippery it is.

They installed a plate on a floor to measure the force exerted by a foot stepping on it and then pushing off. Next, a banana peel was set on the floor, with the outer surface facing up and the inner (and slipperier) surface facing down. It turned out the peel was extraordinarily slippery—five times slipperier than leather on wood. A foot slipping on a banana peel was not that different from a ski sliding on snow.

But the Japanese scientists didn't stop there. They compared the banana peel with other fruit waste, including apple peels, tangerine skins, and the peel of the lemonlike citron—but none of these even came close. (These results don't explain why slipping on orange peels was a comedy routine even before banana peels.)

This experiment goes some way toward clarifying why bananas have become synonymous with slipperiness, but the researchers wanted to know more. They noted that the inner surface of a banana peel is coated with a water-rich gel, probably a protein-sugar mix. They concluded that this gel was responsible for the slipperiness because a peel that has been squished underfoot is much less slippery. The water has been pressed out and the structure of the gel has changed.

Whee!

Today, the banana peel gag has all but disappeared from comedy routines. It could be that it just ran its course—after all, how many banana pratfalls will make you laugh? Also, we no longer take the same casual approach to street garbage that we once did. It's rare today to see a discarded banana peel on the sidewalk. But there's also another possibility: a different banana.

Bananas are a multibillion-dollar global business. For many countries, they are a staple food: China and India don't even export any of their huge banana crops. But for most Latin American countries, bananas are their first or second most important export. In the early twentieth century, companies seeking a monopoly in the banana business preferred to cultivate a single variety of the fruit. A variety called the Gros Michel (the Big Mike) was the chosen one. These bananas were thick-skinned and didn't bruise easily. The bunches were tight and usually contained several bananas. And it took them a long time to ripen, reducing the risk of rot if a shipment was delayed.

Another huge benefit for the banana producers was that the plants were clones—meaning that wherever Gros Michel bananas were grown, they were genetically identical. Every banana in the supermarket was an identical twin to every other. This provided a predictable and easy-to-grow banana, but it also created a risk: if an infectious agent is introduced to a crop like this, which has no genetic variation, the pathogen is likely to run wild. That's exactly what happened to the Gros Michel.

In 1903, a fungus called Panama disease race 1 struck banana plantations throughout Latin America. It spread slowly but inexorably over the next fifty years and eventually wiped out the Gros Michel plantations. That variety is no longer grown in Latin America, the source of almost all the bananas eaten in North America. (You can still get Gros Michel bananas in Asia.)

Very fortunately, there was a banana variety called the Cavendish that proved resistant to Panama disease. If you have bananas in your kitchen right now, they're almost certainly Cavendish. And that has allowed the Canadian and American love affair with the banana to continued unabated—to the tune of 12 billion bananas a year.

Gros Michel   Cavendish

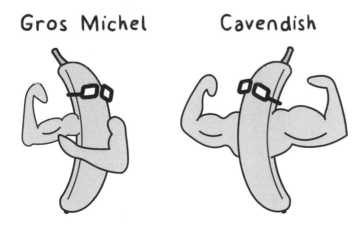

In some ways, though, today's situation is eerily similar to that of a century ago. All bananas grown for export on vast Latin American plantations are Cavendish. But as we now know, maintaining a single variety of banana in a world where fungi are always mutating and evolving is an agricultural risk. When a fungus travels easily, sticks tenaciously to clothing and equipment, and remains infectious in soil for at least ten years, it's almost impossible to prevent its spread.

Sure enough, there is now a new version of Panama disease, called Panama disease race 4. This new fungus destroys Cavendish banana plants and has been spreading in Asia and Africa, and in late summer 2019, it was discovered in banana plantations in Colombia. It's almost a certainty that it will spread throughout Latin America and endanger the entire banana crop there.

Science _Fact!_   _Why does artificial banana flavor never taste like real bananas? It's possible that it does taste like bananas—just not the ones we're familiar with. Banana oil, a flavoring that contains chemicals found in bananas, has been around since the late nineteenth century, long before the Cavendish and even the Gros Michel. So it's possible that it simply tastes like a variety we don't recognize._

There is a ray of hope, though: an international team of scientists genetically engineering several types of Cavendish bananas has found at least two that have remained resistant to Panama disease race 4. If these results hold up, perhaps the new varieties can be planted rapidly enough to keep pace with the spread of the fungus. But the Cavendish banana, whether genetically altered or not, is still a single variety and will therefore be susceptible to future fungal diseases.

It's said that the Gros Michel was creamier and perhaps a little sweeter than the Cavendish, but I know of no attempts to compare the slipperiness of the two. Maybe the disappearance of the Gros Michel is the real reason the banana peel gag fell out of fashion?

History Mystery

Does the Loch Ness Monster really exist?

It's always sad to announce the extinction of an animal, especially an iconic one. But I think we have to accept that as the fate of the Loch Ness Monster. Perhaps the pain of loss is alleviated somewhat by the fact that it never existed in the first place.

Give the legend credit for staying power, though: it dates back more than a thousand years, and throughout the twentieth century, there were almost too many "sightings" to count. Yet, in the end, there is absolutely no evidence that

Loch! Can't ya see me?

anything like a monster is living or has lived in Scotland's Loch Ness. There are still some true believers, but the rest of us can close the book on it.

Nessie was supposed to be a monster indeed—a huge dinosaurian creature with a long neck and tail that was seen both on land and swimming in the waters of the lake. As places to conceal a large creature go, Loch Ness is just about perfect: its surface area is 56 square kilometers (22 square miles), and it is 230 meters (755 feet) at its deepest point. Also, the peat-filled water is extremely murky. If you saw something strange there once, it's quite possible you, or anyone else, would never see it again.

The first mention of Nessie dates from the year 565, when an Irish missionary named St. Columba repelled a monster threatening a swimmer—an act that was deemed a miracle. But most of the excitement around Nessie really began in the 1930s. There's a fantastic 1933 report in the *Inverness Courier*, describing a sighting by a local couple of something strange less than a mile offshore. "The creature disported itself," the newspaper recounted, "rolling and plunging for fully a minute, its body resembling that of a whale, and the water cascading and churning like a simmering cauldron. Soon, however, it disappeared in a boiling mass of foam. The beast, in taking the final plunge, sent out waves that were big enough to have been caused by a passing steamer." That would have been a frightening sight!

Loch Ness Monster reports peaked in the 1930s. The most famous image—known as the "surgeon's photograph"—was published in 1934. The man who took the picture had wanted to remain anonymous but was soon identified as a London doctor named Robert Kenneth Wilson. His name is now forever linked to the legend.

Countless other sightings soon followed, along with more photographs and even sonar recordings from the depths of the lake. Sonar is able to detect the presence of moving objects underwater, which is key in Loch Ness, where the visibility is minimal. Several sonar recordings have seemed to show the presence in the lake of something bigger than the average fish, although it's difficult to differentiate a single object from, say, a school of fish. I have talked to a scientist who was on board a boat when suddenly sonar revealed a large object moving underneath them. The scientist raced to the deck but saw no evidence at the surface of anything at all.

Science Fiction! In 1975, British naturalist Sir Peter Scott, writing in the science journal Nature, proposed a scientific name for the Loch Ness Monster: Nessiteras rhombopteryx, which translates to "Ness inhabitant with diamond-shaped fins." It sounded like a scientific name for a species, but a Scottish politician immediately noticed that the new name was an anagram for "monster hoax by Sir Peter S."

If there were a creature in the loch, what might it be? Probably the most popular suggestion is that it's a plesiosaur, a large long-necked marine reptile that has been extinct for 80 million years. But somehow, the argument goes, a few of these relics from the age of the dinosaurs have managed to survive in the lake. This is beyond unlikely.

The Greenland shark is another candidate. This species has the size to qualify as a monster: it can grow to more than 6 meters (20 feet) in length. It is found in the waters of the North Atlantic and occasionally penetrates deep into fjords, persuading believers that some traveled to the loch all the way from the North Sea.

Other candidates have included trees, seismic gas, an otter, and even several otters.

I am Nessie.

One of the most frequently mentioned possibilities is that Nessie is a giant eel, and recently eels have reentered the Loch Ness picture. In 2019, a team of researchers from the University of Otago in New Zealand sampled the lake waters for DNA. While they found no evidence of a prehistoric reptile or a giant shark, they did find lots of eel DNA. While the leader of the expedition, Dr. Neil Gemmell, was careful to say he couldn't discount the possibility that people might have seen a giant eel in the loch, he admitted that the main reason for his expedition was to promote DNA research. Besides, the largest eels top out at around 5 kilograms (11 pounds). The most recent claim was that Nessie is a giant catfish. There was a photo too—unfortunately, it was of a fish caught in Italy.

What about the classic "surgeon's photograph"? It's true that it showed the head and neck of a large animal extending out of the water, and for many years it was the most prominent piece of evidence for Nessie. But smart detective work in the 1990s revealed that the public version of the photo was the cropped version of a larger original photo. The cropping made it difficult to get a sense of scale. When an analysis of the entire photograph was performed, it revealed that what looked like waves were actually ripples, and the "gigantic" animal was actually only 60 to 90 centimeters (2 to 3 feet) long. Sadly, the photograph was an elaborate hoax dreamed up by a small group of pranksters, including Dr. Wilson. They used a toy submarine for the body, fashioned a head and neck, and snapped a picture that helped create a legend.

So it has come to this: There is no reliable evidence of a large unknown creature in Loch Ness (even though sightings are still reported every year). Instead, we have scientists fishing for DNA to promote their own research. Time to declare the poor beast extinct.

Part 4
Perplexing Phenomena

What is a Mexican jumping bean?

I USED TO WONDER IF MEXICAN JUMPING BEANS really existed or were just something I'd seen advertised in comics. But they are a real thing. The town of Álamos, Mexico is famed for its harvest of jumping beans, and if you want some for yourself, you can order them online. A company called Chaparral Novelties has sold 3–5 million of them a year for at least the past three decades.

That's the business side, but here's the biology: A Mexican jumping bean is actually a seed with a moth larva inside. It jumps because the larva twists and turns, and these movements set the seed in motion. How does the larva get inside the seed? It all starts in the spring, when a desert shrub known as *yerba de flecha* (arrow herb) displays its flowers. Female moths lay their eggs in the flowers, and those flowers eventually develop into seedpods, trapping the larvae inside. In time, the seeds fall off the plant to the ground below.

You seem jumpier today.

It's my larva.

Yes, the larva is trapped inside the hard-shelled seed, but it's protected, too—and fed, because the larva gnaws away at the inside of the seed. As long as everything goes well, the larva can spend months inside the seed until its transformation into an adult moth is complete. The problem is, the larva won't survive if it's constantly exposed to the hot Mexican sun. It risks dehydration or death. This is apparently why the "bean" "jumps"—the larva is seeking cooler temperatures.

How does all this work? Well, if the ground starts to heat up, the larva becomes extremely active. Every time it twists and turns, it sets the seed moving. Some beans have been recorded jerking or flexing once every couple of seconds or even faster.

Now, surely a fat grub sitting blindly inside a seed can't be steering the seed in the direction it wants? Actually, it can. The larva anchors itself to the walls of the seed with silklike threads, to ensure that its movements are transferred to the seed itself. If the larva scrambles around inside, tugging on the threads as it goes, the seed begins to roll. If it strikes a wall, the seed jumps. And because the larva is able to sense heat, it will maintain these erratic movements until it finds a cooler place.

The shape of the seed helps the larva get it rolling, but it also limits how far it can travel. The seed has three sides, two flat, the third round. When a moving seed stops, it can come to rest on either one of the flat sides or balance on the curved side. The larva takes advantage of this irregular shape by rolling the seed, which gains distance; jumping, where the seed launches into the air and comes to rest on the same face; and flipping, where the seed becomes airborne and lands on a different side. Flipping is the most dramatic: it can move the bean as much as

4 centimeters (about 1½ inches). For the most part, though, the bean's movements are short-range—a roll here, a flip there.

The larva spends the next few several weeks like this, moving the seed around while gradually eating away at it from the inside. Just before it pupates (and becomes inactive), it gnaws a hole in the side of the seed and then seals it back up with a patch of thread. In the spring, the adult moth finally hatches from the pupa, follows a path of silk to the exit hole, and flies off into the world to mate—it will live only a couple of days.

Did You Know . . . There are several living things other than jumping beans that move by rolling. These include the web-toed salamander, the Namib golden wheel spider, and the wood louse. The salamander curls into a ball and can roll down hills. The spider escapes wasps attacking it by tumbling down sand dunes at twenty revolutions per minute. And the wood louse curls up to protect itself, and if it's on a slope, it starts to roll. But none of these animals need to roll—they all have legs they can use instead. The one organism that stakes its life on rolling is the tumbleweed. When its seeds are ready, this mostly stationary plant detaches itself from the ground and rolls along, distributing its seeds everywhere it goes.

It's bizarre to think that larvae encased in seedpods could become a novelty item sold in toy stores, but what's even weirder is that they've inspired engineers who work with robots. David Hu and his colleagues at the Georgia Institute of Technology looked at the movements of Mexican jumping beans and wondered whether that approach could be used with robots—specifically, robots that roll.

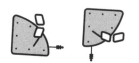

Rolling robots—think of them as metallic spheres—should have some advantages over other designs. Not only are they are encased in a protective metal shell, but rolling is the most efficient way to move over flat surfaces. (The lack of flat surfaces in nature is one of the reasons there are no animals that use wheels—they don't adapt well to varied terrain.) The engineers were impressed with the Mexican jumping beans they studied. They concluded that the beans' shape and movements combine to allow them to travel in one direction, sense the temperature, and either turn or keep moving.

Science Fiction!   *The hoop snake is said to grasp its tail in its mouth and somersault after prey, unrolling at the last second to deliver a killing shot of venom. It doesn't exist.*

Also, because the seed is irregularly shaped, it can climb over much higher objects than it could if it was spherical, from a mere 0.02 millimeters to 1 centimeter, an incredible 500-fold difference. The engineers pointed out that if robots were shaped like jumping beans instead of being perfectly round, they might have similar abilities.

Hu and his teammates designed software that could mimic the jumping bean's behavior and programmed an off-the-shelf robot to act exactly that way. They suggest that small spy robots modeled after jumping beans could be programmed to avoid light instead of heat and scurry for the shadows if a lamp is turned on.

Why is your mirror image backward?

WE ALL KNOW THAT OUR IMAGE in the mirror isn't exactly like the real us. It looks *a lot* like us, but tiny details are amiss. That freckle on your right cheek? It's on the left of your reflection. Your left eye is your reflection's right eye, and the print on your T-shirt is backward. In fact, everything left has been switched to the right and vice versa. The mirror is doing the switching—or so it appears.

A mirror is an unusually flat surface made of glass with a backing of silver. When light bounces off a mirror, it doesn't scatter off in all directions—instead, the angle of incidence will equal the angle of reflection. In other words, if light strikes the mirror at a 60-degree angle, it'll reflect back at 60 degrees, only in the opposite direction. Picture yourself throwing a ball at an angle toward the pavement—it will bounce away at the same angle.

It's all a matter of perspective.

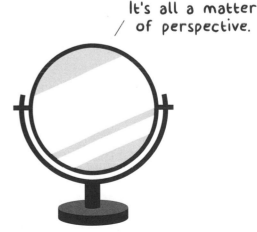

149

When you're staring into the eyes of your reflection, however, you are seeing light that traveled to the mirror and straight back. The flight of the photons is so true that your reflection is precisely like you—except for that right-left issue. The reflection's right thumb is a perfect image of your left thumb, but it's not built like a left thumb. You couldn't superimpose the two.

Flight of the Photons

One way of visualizing this difference is to imagine you are wearing a glove on your left hand. If you peel it off so it's inside out and then hand it through the mirror to your reflection, the mirror version of you can put the glove on its *right* hand, where it fits perfectly. You actually turned the glove into a mirror image by pulling it inside out. And if the mirror then handed the inside-out left glove back to you, it would now fit your own right hand.

 TRY THIS AT HOME! Write a word like "book" on a piece of paper using a thick dark marker. Now hold the paper up to the mirror—you'll see the word "book" written backward. If you then sneak a look at the back of the paper, you'll see exactly the same thing: ʞood. And if you write "book" backward on a piece of paper, you'll see it appear correctly in the mirror and through the back of the paper as well. At no time has right been switched with left. If they had, the letters would appear in the right order in the mirror; instead, they've been flipped like the glove.

How does a right hand or eye or foot get turned into a left? One way is simply to flip everything from right to left. But then your right hand, as it flipped, would become the right hand of your reflection. That isn't what happens with a mirror, as we saw with the glove example. Your right hand became the left hand of your reflection, not the right. The glove was turned inside out, not flipped.

If the mirror flipped you from left to right, your image would be the same one you see in every selfie. But the you in the selfie is not the same as the you in the mirror. Some of you may be sorry to hear this, but the you in the selfie *is* the way you look to others. The you in the mirror is not the actual you—it's you turned inside out.

Confusing? Try to imagine a perfect mold of your entire body being carefully peeled off and then held out in front of you. As long as it maintained its shape perfectly, it would be the exact equivalent to your mirror image.

Science _Fact!_ *When you have an itch on your arm, you scratch. But mirrors have been used to demonstrate that sometimes even if the wrong arm is scratched, the itchy person will still feel better. In this experiment, a mirror is placed between a subject's forearms, with the mirrored side facing the left. The subject is positioned in such a way that she will see only the reflection and think that's her right arm. (In other words, her left arm isn't visible to her.) Next, a small amount of histamine is injected into the actual right arm, creating a serious itch. If the subject sees her left (non-itchy) arm in the mirror being scratched, she'll think it's her itchy right arm, and the scratching will make the itch subside! This test shows that itchiness is really perceived in your brain, not on your skin, and if your brain is convinced that an itch is being scratched, the intensity will lessen.*

Even though our mirror image is not the way others see us, we know it's us in the mirror staring back. But this isn't true for every species. The "mirror test" is commonly used to assess self-awareness in animals. This test was first applied to chimpanzees. A chimp is briefly anesthetized, then a researcher puts a red dot on its forehead. When the animal awakens, it is allowed to look at itself in a mirror. If it knows the reflection is itself—in other words, if it has self-awareness—it will be curious about the red dot and poke at it or even try to remove it. Other animals may look in the mirror and see the dot, but they don't try to remove it because they don't realize the reflection is theirs.

Animals that seem to recognize themselves in the mirror test include the great apes, dolphins, and orcas. Pandas, sea lions, and several monkey species have failed. There's even evidence that ants behave differently in front of a mirror. Surprisingly, the superintelligent gray parrot has failed the test, while the magpie—which is smart, but not quite parrot-level smart—has passed.

This isn't me.

Humans figure out mirror reflections at about eighteen months of age. If you review the list of animals that pass the test, it's tempting to equate mirror self-recognition with intelligence. But some scientists argue that all it really tests is whether an animal cares about the appearance of its body. And some, like dogs, don't depend so much on vision. What we need for them is a "smell mirror" that reflects their odors back to them.

What is the Doppler effect?

You know the Doppler effect even if you don't remember its name. It's the sound of something rushing past—an ambulance siren, a train whistle. As the sound approaches you, it has a higher pitch. As it recedes into the distance, the pitch drops. In other words, its frequency changes. You hear the note as it was intended to be heard only when it's right beside you.

This effect is named after Christian Doppler, the Austrian physicist who first described it. Funnily enough, he was thinking not of sound but of light. He believed that the color of a star (the frequency of its light) would change according to the star's speed in relation to Earth, much as the frequency or pitch of sound from a moving vehicle drops as it passes by. He was thinking particularly of binary stars, where one star, as it orbits the other, sometimes approaches us, sometimes recedes. And he was right this would happen—anything that produces waves can create a Doppler effect.

Even though Doppler hit on this idea while studying the light from stars, we commonly experience it with sound. The classic demonstration of the sound version of the Doppler effect was performed in 1845. An open train car full of trumpet players "rushed" past a railroad station at 65 kilometers (40 miles) an hour. The trumpeters were all playing the same note. At the same time, there were other trumpeters inside the station, playing the same note for comparison. As the train approached and then passed, the pitch of the onboard trumpeters' note changed, from a little higher as the train neared to a little lower as it sped away.

Did You Know . . . The experiment with the trumpeters on the train worked at a relatively slow speed, but the difference in pitch between the moving and stationary notes was only about the same as the difference between adjacent notes on a piano. What if the train moved so fast that as it approached, the note was above the upper range of human hearing, and after it passed, the note was below? If you were standing at the station, you'd see the train coming but hear nothing, and then as it swept past, you'd perceive a high-pitched note falling wildly before disappearing again—all in a second or so. Unfortunately, this experiment might be hard to pull off. As it approached the station, the train would have to be traveling around the speed of sound, which is 1,225 kilometers (761 miles) per hour, but when it passed, it would have to hit something like twenty-one times the speed of sound, or 25,600 kilometers (16,000 miles) per hour. Even if that were possible—it isn't!—the sonic boom would wipe out any Doppler effect.

The physics of the Doppler effect is pretty straightforward. When you press a car horn, it emits a series of sound waves, spaced equally apart and traveling at the speed of sound (of course!). When the waves arrive at your ears, your brain interprets them as a sound. Because the waves emitted by the horn are always the same wavelength, the space between them is consistent. But when the car is approaching you, it gains a little ground between each wave, and because of that, they get pushed closer together. There are more waves per second, so the sound has a higher frequency. As soon as the car passes, the opposite happens: it moves slightly farther away between each wave, so they're no longer crowded together and the sound drops in pitch.

 TRY THIS AT HOME! You can try to generate the Doppler effect yourself if you have a tuning fork or something else that emits a steady pitch. Hold the device at arm's length straight out at a friend, then pull it back and extend it again as fast as you can. Your friend might be able to hear a slight change in pitch. You do have to move your arm pretty fast, though—this is not an exercise for those with rotator cuff injuries!

But the Doppler effect is not just with sound. It also affects light—which is what Doppler himself claimed. He argued that you should see the effect when two stars orbit each other. (He was right about that, but it's too small an effect to account for the colors of the stars he was studying.) As one star approaches us, its waves are closer together—just like those of the car horn—and that shifts its light slightly toward the blue end of the visible spectrum. When the star is retreating, the waves are more spaced out and shift toward the red.

On a much grander scale the Doppler effect tells us that the universe is expanding and galaxies are rushing away from one another. We know that because the light from them is shifted from the blue end of the spectrum toward the red. The faster they're receding, the redder the light appears to us.

There are undoubtedly millions of distant galaxies shining in visible light that we'll never see, at least with the naked eye, because their light waves have been stretched out so much by the expanding universe their light has been shifted into the infrared, which is invisible to us (although not to our telescopes).

Even ultrasound waves can be Doppler shifted, and that can be used to measure the speed and direction of blood flow in the liver or the heart. The ultrasound waves bounce off objects like red blood cells and return to the device with slightly different frequencies, indicating movement either toward or away from the ultrasound source. It's a noninvasive way of determining if blood is flowing in, say, an artery, and if so, in which direction and how fast.

But perhaps the most common use of the Doppler effect is in radar: it will tell a police officer how much over the speed limit you're going or a pitcher how fast the baseball crossed the plate. The difference is that the policeman takes a bead on you when you're approaching and so records a higher frequency of reflected waves, while the baseball radar gun can accurately measure both the speed of the pitch as it approaches and the speed of the ball when it leaves the bat.

Do all snowflakes have six sides?

Snow is amazing stuff. Snowy landscapes can be beautiful, and snowflakes are sensational geometric creations. Most of us learn in kindergarten that snowflakes have six sides, or arms, and that no two flakes are identical, but we don't usually go much deeper than that. That's unfortunate because the cool stuff is just around the corner.

So yes, all snowflakes do have six sides, but why? It starts with H2O, water—two hydrogen atoms and one oxygen bonded together. In outline, a water molecule looks a little like Mickey Mouse's head: one larger circle (Mickey's head, the oxygen atom) with two smaller ones perched on top (Mickey's ears, the hydrogen atoms). When water is in liquid form, its molecules pretty much fill any empty space among them, easily slipping past

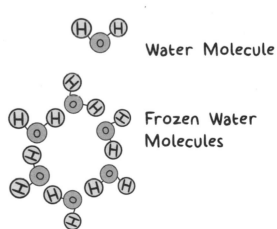

Water Molecule

Frozen Water Molecules

each other like jelly beans in a jar. The water molecules are warm and have lots of energy—they're not going to stick together for more than a moment.

But when the temperature drops, the chemical attraction between water molecules overcomes their ability to move, and they start to cling to one another. Eventually, they lock into a rigid six-sided arrangement, a hexagon, with one water molecule at each corner. This is the most efficient way they have of forming stable bonds. There's now a lot of space in the middle of the hexagon where before there was none. That means ice takes up more volume than water, which is why ice cubes float.

This is the all-important pattern, and while I said these hexagons are rigid, the truth is that nothing in the atomic world is. A water molecule in a six-sided crystal is likely to jump out of position roughly every millionth of a second, wander around a bit, then kick another molecule out of a different crystal and take its place. But as long as the temperature stays low, the overall structure remains stable. Those six water molecules are the foundation of a snowflake. As they're joined by others, the snowflake grows. But no matter how big it gets, it maintains its overall six-sided geometry.

Because each flake has a unique trip down through the atmosphere—through varying temperatures, wind, and humidity—there is no such thing as a standard snowflake. They're similar because they're all six-sided, but the resemblance ends there. Snowflakes are shaped like columns, needles, prisms, or the familiar six-armed version.

Science Fiction!   We're all told that no two snowflakes are the same, but is that true? Maybe not. In 1988 cloud scientists flying at about 6,000 meters (20,000 feet) snapped a picture of what appear to be two identical flakes. They look amazingly alike, but there is only one photo, from one angle, so skeptics can always argue, "If you just turned it over . . ." This is the only pretty solid claim of two identical snowflakes.

Once all those snowflakes reach the ground, they will either melt immediately or start to clump together, crystals filling the empty spaces, while the whole snowpack begins to compact. Flakes that were separate are soon close enough to make contact. This is especially likely to happen if temperatures are temporarily over the freezing mark—or just high enough to melt some of the snow—then drop again to refreeze it. That refreezing promotes something called sintering, or ice welding onto ice, creating uncountable numbers of tiny ice bridges that solidify the mound of snow. This is the same process that causes ice cubes to clump together if you leave them in the bottom of the bucket too long.

TRY THIS AT HOME! If you want to make the perfect snowball, the key is to wait until it's exactly the right temperature outside. Ideally, you want to exert enough pressure when you squeeze the snow together that the upper surface melts and then refreezes as soon as you stop squeezing. If it's too cold, you can't exert the necessary pressure, the melting-and-refreezing process doesn't happen, and your snowball falls apart.

If you've ever walked on snow that has had time to pack together like this, especially on a very cold day, you will hear it squeaking underfoot, a sound unlike snow at any other temperature. The temperature is crucial: There's a threshold of around −10°C (14°F). Above that temperature, snow makes little noise—at most a soft swish or a gentle crunch. But below that mark (and especially when it's really cold, like −30°C, or −22°F), the sound is legitimately a squeak. It's not a pure musical note, but a chaotic mix of different frequencies played out in a fraction of a second.

What's happening is that at higher temperatures the pressure of your weight will melt the snow underfoot, immediately dampening both the impact of your step and the sound of the snow. But if it's cold enough and there's no melting, the impact of a footstep is much greater. In the bitter cold, crystals that were sintered, or welded together, will shatter instead of melt. Could the simultaneous breaking of trillions of such bonds be responsible for the squeak? That's the best explanation we have so far.

You've probably also noticed the opposite effect: as snow is falling and piling up on the ground, the surroundings seem to quiet. When snow first lands on the ground, it is fluffy and full of air, much like common soundproofing materials, which work by providing spaces for sounds to get lost in. When sounds strike snow, it's the same story: they penetrate the snow but don't emerge. The layer of fresh snow on the ground creates the hush of a new snowfall.

Can a butterfly flapping its wings in Brazil set off a tornado in Texas?

THE ASTOUNDING IDEA that a tiny butterfly could, merely by flapping its wings, cause a tornado thousands of kilometers away comes from Ed Lorenz, a scientist famous for his contributions to chaos theory, the branch of mathematics describing events that play out in unpredictable ways. The weather is a perfect example: the forecast for tomorrow is usually pretty accurate, but the two-week forecast is much harder to get right because so many factors can influence weather systems between now and then. Lorenz deserves credit for explaining why that is.

It's the butterfly effect.

Of course, the weather's not the only thing that's difficult to forecast. There is a famous proverb that describes how seemingly unimportant things can have great and unexpected consequences. It's centuries old and exists in several different versions, but one goes like this:

For want of a nail the shoe was lost.
For want of a shoe the horse was lost.
For want of a horse the battle was lost.
For the failure of battle the kingdom was lost.
All for the want of a horseshoe nail.

No rider would ever pick up a lost nail from a horseshoe and think, "Uh-oh, the kingdom is going to fall!" That chain of events could only be understood in hindsight. In the same way you could never watch a butterfly flap its wings and *predict* that it would cause a storm somewhere far away. Nor, as Lorenz pointed out, could you claim that the butterfly's flap would do the opposite and prevent a distant tornado. And that is really the point: the chain of events is too long and too complex. We can never precisely identify all the factors that end up producing a tornado, or any other weather-related event, because many of those factors are so minuscule they're untraceable. When a huge storm hits, we may suspect that some sort of microturbulence from days or even weeks before was to blame, but we'll never know for sure.

The difference between tomorrow's forecast and the forecast for two weeks from now is that tiny effects take time to manifest. An example from one of Lorenz's experiments illustrates this point. He was running a weather program on a computer, using twelve numbers describing atmospheric conditions such as temperature to try to predict the weather ahead. At one point, he wanted to get finer detail from the computer, so he stopped and reset it, then started it again with the number 0.506, instead of the original 0.506127. The new shortened number was identical to the third decimal place, but different from there on—a seemingly tiny difference.

Lorenz then went for coffee, and when he returned an hour later, he found that the computer program was producing a completely different forecast. He saw that as a result of the slightly less refined number, the weather prediction had begun to diverge from the original, generating a widening disparity that doubled every four days. By the time the prediction was in its second month, it was utterly different.

It's going to be buggy out there today.

Did You Know . . . In 1952, science fiction author Ray Bradbury wrote a short story called "A Sound of Thunder," about a man named Eckels who takes a time machine back to the age of the dinosaurs to kill a *T. rex*. When it appears, the *T. rex* is absolutely terrifying, and Eckels loses his nerve. He runs back to the time machine, letting others kill it instead. But in his panicked retreat, he steps off the designated path, which is meant to prevent hunters from standing on the soil and inadvertently killing something, altering the future. When Eckels and the others arrive back in the present, things have changed. And as he takes off his shoes, he finds, in the soil crushed underfoot, a dead butterfly.

Could a butterfly actually play the role of weather disturber? Most scientists doubt that. For one thing, the disturbance of the air generated by the flapping wings is enough to get the butterfly aloft, but it's infinitesimally small compared to the other breezes and gusts of wind that surround it. It would be drowned out in seconds. It also seems unlikely that a butterfly in Brazil could influence the weather in Texas because the two places are in different hemispheres and the respective weather systems largely keep to themselves.

Could some other apparently trivial effect play the role of Lorenz's butterfly? Some have suggested that the shape, size, and movements of individual clouds could, but at the moment those are far beyond the capabilities of weather forecasters' computers.

None of this takes away from the importance of Lorenz's work, though. His core findings—that tiny and likely unidentifiable factors can cause a series of events to unwind in ways no one anticipated—have become a key piece of chaos theory and have led to new ways of thinking about modeling predictions by computer, especially when it comes to the weather.

All of this might make you wonder how climatologists can predict the future effects of climate change. If it's impossible to be perfectly accurate about the weather weeks from now, how can they say anything about what will happen ten or more years into the future?

There's a crucial difference here. In the short term, weather develops in response to those tiny initial influences that set it off on certain paths. But in the much-longer term—years, not days—the atmosphere settles down into stable and consistent patterns, responding not to short-term influences but to long-term trends like temperature. Weather describes the unique set of steps you take on your everyday hike, never exactly the same twice, but the climate describes where you will be when you reach your destination. Neither can be 100 percent accurate, but the latter is easier to predict.

The answer to this question, then, is no, a butterfly flapping its wings in Brazil cannot set off a tornado in Texas. But the principle behind the question is true: small, undetectable events can alter what happens on a much larger scale. And that is a powerful scientific idea.

I always fly south to
avoid the winter weather.

What is the Moral Machine experiment?

THE MORAL MACHINE IS AN ADAPTATION of a thought experiment known as the trolley problem, so let's start there. Imagine there's a trolley hurtling down some tracks. Just ahead, there are five people tied to the rails. If the trolley is allowed to continue, those people will die. But you are standing next to a switch, and if you throw it, the trolley will be diverted onto a spur line, where there is just a single person who will die. If you do nothing, five die. If you take action, you save five but kill another. What do you do?

There are different varieties of the trolley problem—one first published more than a hundred years ago had a railwayman controlling the switch, but the single person on the spur line was his son. I'm sure you can appreciate that ethically there's much

I'm morally corrupt.

ERROR!

to discuss here. For example, if you pull the switch to save five lives, you're choosing to kill someone and acting to make that happen. Leaving the switch alone at least means you aren't a participant. The Moral Machine experiment has these same kinds of ethical issues, but it adds some technological ones as well.

The Moral Machine is an online platform—basically a game—created by researchers at MIT. It's intended to challenge players with ethical issues just as the trolley problem does, but the ultimate goal is to find out how self-driving cars should be equipped to make similar decisions. By gathering responses from players in more than two hundred countries, the Moral Machine has sampled what those players would like a self-driving car to do when faced with a difficult choice.

Self-driving cars, though still controversial and imperfect, will likely play an important role in the future. Once in use they should eliminate accidents caused by distracted driving, drunk driving, or road rage. They can negotiate city streets and even respond to emergency situations because they are equipped with artificial intelligence (AI). They need extremely powerful processing capabilities to do all this, though, because of the amount of information they must take in from sensors and the instantaneous decisions they must make. The decision part is tricky: the AI system needs to be trained to react appropriately in situations it's never before seen. A self-driving car is not preprogrammed: it has to rely on its own smarts.

These cars are going to have to make quick decisions in pressure situations. For example, many North American drivers are accustomed to encountering deer and moose on the roads, and there is an informal rule that you don't swerve to avoid the deer because you'll likely end up in the ditch, injured or dead, but you do swerve to avoid the moose because colliding with that enormous animal is even more dangerous than driving into the ditch. Self-driving cars will be perfectly able to tell a moose from a deer—quicker than we can—so this scenario might not present a dilemma. In many other situations, though, a self-driving car could find itself making a decision that risks causing death. How would you prepare it for that?

That's what the creators of the Moral Machine wanted to know, so they built a game that confronted people with accident scenarios and asked them to choose who'd live and who'd die. Millions of people around the world have voiced their opinions in this more elaborate version of the trolley problem.

Overall, people who've played the game would save a woman with a stroller over anyone else, closely followed by a girl, a boy, and a pregnant woman, in that order. Next in line were doctors (males narrowly favored over females), then athletes and executives (of both genders). Finally, there were those who didn't rank as highly, including old men, old women, dogs, criminals, and, in last place, cats. (The dilemma posed by a cat in a stroller was not addressed.)

So given the choice between hitting a dog or a cat, most people would want their self-driving car to save the dog. Any athlete would be allowed to live in preference to any old person. A male athlete versus a female athlete would be a coin toss.

But the Moral Machine also gathered data on who was responding and where those people lived, and this allowed the designers to demonstrate differences in choices between countries. And some of those differences were significant.

The researchers identified three clusters of countries: the western cluster, made up of North America and several European countries; the eastern cluster, which was Asian countries; and the southern cluster, made up mostly of Central and South American countries. All three clusters differed, sometimes in major ways. The western cluster, for instance, valued car passengers over pedestrians more than the other two. The eastern cluster, which includes countries like Japan and China, where there has always been a reverence for age, was not nearly as pro-youth. The southern cluster had the strongest preference for saving women and people who are fit. Cultural differences were also intriguing. In many affluent countries, for example, there was a preference for saving a pedestrian crossing the street legally over one who was jaywalking, but in poorer countries there was much more tolerance toward the jaywalker.

This is a fascinating survey, and with 40 million entries, the data will be mined for a long time to come. Of course, a computer program can't perfectly simulate real life. If you've ever been in a collision, you know you have no time to think about what you're going to do—you react unconsciously and automatically. But self-driving cars will be superior to us in many ways, and perhaps when put in these either/or dilemmas, they'll be capable of finding a third option that saves everyone. Regardless, this survey shows that a self-driving car might have to alter its decision-making depending on where it is in the world.

As the builders of the Moral Machine say, "Never in the history of humanity have we allowed a machine to autonomously decide who should live and who should die, in a fraction of a second, without real-time supervision."

What's the best way to shuffle cards?

HAVE YOU EVER WONDERED IF SHUFFLING a deck of cards is doing any good? Is it really mixing the cards thoroughly, randomizing them enough that cheating isn't possible? It depends on how you shuffle.

Cards are shuffled to eliminate the possibility that a predictable order has carried over from one game to the next. There are several ways of shuffling, and each one is effective—if you do it right. The riffle shuffle is probably the most common. This is the one where you split the deck roughly in half, then, with the two halves facing each other, press down on the middle of the cards with your index fingers and use your thumbs to release the cards gradually so that they alternate, like the teeth of a zipper. Just like that, you've (partly) shuffled the cards.

It's unlikely that you've done this perfectly, however, with each card from one half of the deck alternating with a card from the other half. And even if you have, after a single shuffle, there's still some predictable order to the cards that might allow cheating, particularly if the deck was perfectly in order—that is, running from ace to king in each suit—to start. The key is to randomize the cards, and the math suggests that when it comes to the riffle technique, seven shuffles will make the deck as random as possible.

We know a lot about the mathematics of shuffling thanks to Persi Diaconis, a mathematician at Stanford University who left home at the age of thirteen to travel with a magician and learn his craft. Eventually, he returned to school and earned a PhD in mathematical statistics from Harvard. He is an expert on randomization and is the guru of card shuffling.

Diaconis and his colleagues showed that it takes five riffle shuffles before there's a dramatic change in the order of a standard deck, and that you need an additional two shuffles to thoroughly mix the cards. In other words, there's a critical threshold that must be reached to ensure randomness. (Perfect randomness is unlikely, but in most card games, practical randomness is sufficient.)

Diaconis has also shown that perfect riffle shuffling—that is, where the cards are released one at a time and alternate exactly—is not ideal. If you have just ten cards and riffle them perfectly six times, they will return to their original order, as if you had never touched them at all. With an entire deck of fifty-two cards, you need to shuffle them perfectly eight times to return to the original order. So there is some benefit to the fact that we humans don't shuffle like machines.

Diaconis also calculated the cool fact that if a person is asked to guess which card is next as they're turned over one by one from a well-shuffled deck, she will on average get 4.5 cards right. If the deck is not as well shuffled—let's say it's been riffled only three or four times instead of seven—she'll guess right 10 times, on average. So a cheater *could* prosper from a poorly shuffled deck.

Seven is the lucky number for the riffle shuffle, but what about some of the other common shuffling techniques? It turns out that the ideal number of shuffles varies depending on the method you use. Take the overhand shuffle, for example. With this, you start with the entire deck in one hand, then pick up maybe three-quarters of the cards in your free hand and begin returning them to your starting hand in small amounts—say, four or five cards at a time, sometimes on one side of the deck, sometimes the other. The degree of randomization depends on how many shuffles you do and how many cards you replace each time. Nonetheless, this technique is nowhere near as effective as riffling; it would take about ten thousand overhand shuffles to randomize the deck completely!

Some people prefer to use smooshing. With this approach, the deck is spread out across the table and the cards are moved around and mixed together again and again. If this is done continuously for about a minute, the deck will be properly shuffled.

How did he shuffle?

 TRY THIS AT HOME! Sometimes human psychology, not mathematics, dictates the outcome of a game. Take rock paper scissors, for example. In this game, winners tend to repeat the strategy they just used, while losers are more likely to change. You can exploit these tendencies to improve your chances of winning. So if you've lost a round, play the one option that didn't appear in the previous game. If you had paper and were defeated by scissors, play rock (because the winner is more likely to use scissors again). If you won by playing scissors to your opponent's paper, play paper in the next round (because your opponent is more likely to switch to rock). Try it!

While coin tossing isn't exactly analogous, it does raise a similar question: Is a coin toss a truly random event? There are varieties of coin tossing, and an especially important factor is where the coin lands—on the ground, in your hand, or on a table. Catching the coin in the hand is probably fairer than letting it land on the floor, because if it starts to spin on the floor, then the bias of spinning coins comes into play. (It has been calculated that the chances of a coin landing on its edge is 1 in 6,000.) But otherwise it's straightforward. Spinning a coin and waiting for it settle on one side or the other isn't random either because the two sides of the coin are not exactly the same (one is heads, one tails), and most coins have a bias to fall on one side or the other.

Persi Diaconis has also studied coin tossing, and he and his colleagues predicted that people tossing coins might slightly favor the side of the coin that was facing up when it was tossed. So a coin showing heads would more likely end up as heads when it landed. The predicted difference was slight, though: instead of exactly 50:50, it was more like 50.8:49.2. But sometimes math and reality appear to clash. A heroic experiment at Berkeley showed that after 40,000 coin tosses (20,000 each by two students), there appeared to be no such bias.

Our fates are just a roll of the dice.

Science _Fact!_ *Sigmund Freud apparently supported the idea of flipping a coin to make difficult decisions—but only if you examine your feelings after the toss. If you're happy with the outcome, that's likely the right decision for you. If you're unhappy, the coin flip is still telling you which choice to make.*

How black is black?

THIS MAY SOUND LIKE AN ODD QUESTION. Isn't something that's black just that? You might think so, but in fact, there are degrees of blackness. And scientists around the world are racing to create the blackest black possible.

What exactly is black? Technically, it's not a color—it's the *absence* of color. White, on the other hand, is a combination of all the colors of the rainbow: red, orange, yellow, green, blue, indigo, and violet. It's a mix of every wavelength of light in the part of the electromagnetic spectrum we can see.

My life is lacking color.

Something that's white reflects almost all light back to your eye. Colored objects reflect some light and absorb the rest. If you have a pen with red ink, for example, substances in that ink absorb most of the light falling on it but reflect the red back to your eye. Black, however, reflects almost nothing, so you see little to no color at all.

The blackness of a paint or an ink is measured by how much light is captured or absorbed. Tens of thousands of years ago, artists used charcoal, peach pits, and even burnt bones to produce black paints or dyes, but today most black pigments are created by industrial chemical processes. And those pigments are much blacker than the ones used long ago—that is, they reflect less light. Yet even today, nothing is 100 percent black, although we're getting close.

He's really captured the light, hasn't he?

A few years ago, scientists at a British company called Surrey NanoSystems announced they'd created Vantablack, a substance that absorbs an incredible 99.96 percent of the light falling on it. (Normal black paint absorbs only 90 percent or so.) "Vanta" is an abbreviation for "vertically aligned carbon nanotube arrays"—what does that mean?

Carbon atoms can bond together to form structures called buckyballs and buckytubes. Buckyballs are soccer ball–shaped molecules named after Buckminster Fuller, the architect who popularized the geodesic dome (think Spaceship Earth at Epcot). Buckytubes are elongated hollow versions of the same. These tubes are extremely small but very strong. Take a single hair from your head, split it lengthwise ten thousand times, and you have something the size of a carbon nanotube.

Did You Know . . . Vantablack has caused controversy in the art world. Surrey NanoSystems granted artist Anish Kapoor sole rights to use the material in his work. Other artists were predictably angry— so much so that Stuart Semple created a pigment he called the Pinkest Pink and announced that he wouldn't sell it to anyone who might give it to Kapoor.

But the blackest blacks could soon be seen everywhere. BMW unveiled a car painted with Vantablack at the Frankfurt auto show, and a Swiss watch company already sells an ultra-expensive timepiece with a Vantablack face.

Now imagine a forest of these, standing upright like trees without limbs—that's a sheet of Vantablack. Shine light on it and the particles of light, or the photons, bounce back and forth among the trees, losing a little energy each time, until they've lost so much energy they've entered the infrared realm and we can't see them anymore. But a few particles are reflected back while they're still within the range of visible light, which is why Vantablack is 0.04 percent shy of being perfectly black.

TRY THIS AT HOME! Although black materials absorb the light falling on them, the light energy doesn't just disappear. Instead, it's converted to heat, and that makes the material very slightly warmer. This is similar to what happens when you close the refrigerator door. The bulb inside turns off, but the light that's already in there is trapped. Where does it go? It gets absorbed by the walls and the food, making all of them just a tiny bit warmer (so little you'd barely be able to measure it).

If you don't believe the fridge light goes off when you close the door, put your smartphone in the fridge with the video recording, close the door, wait a second or two, then open the door again and see what the video shows.

Recently, Vantablack lost its title as the blackest black. In late 2019, engineers at the Massachusetts Institute of Technology tweaked the method of producing standing nanotubes and created a material that absorbs 99.995 percent of incoming light, making it ten times blacker than Vantablack (although it still leaks an extremely tiny amount of light). Looking at a sheet of this material is like looking into a bottomless pit or a black hole. Even if there's a bump or a ridge on the surface, you can't see it.

These ultrablack materials might be useful for astronomy. For example, they could absorb any glare finding its way into a telescope and obscuring whatever celestial body the astronomer is investigating.

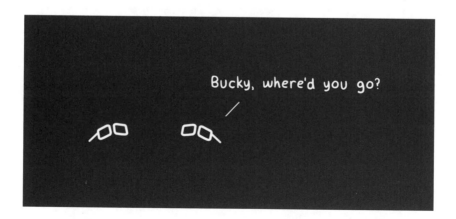

Science _Fact!_ And what about white? Yes, there are advances there, too, although they're perhaps not quite as spectacular as those in the race to the blackest black. European scientists have borrowed a trick from a Southeast Asian beetle whose shell is extremely white (probably to camouflage it among white fungi). The beetle achieves this with layers of a substance called chitin—the stuff that insects' exoskeletons are made of. Chitin is only a tenth as thick as a human hair, and it's structured in a way that scatters incoming light rather than absorbs it. Because colors bounce around and reemerge, almost none of the light is lost—and the result is a brilliant white. The whiteness is even more amazing given that the foundation layer of the beetle's shell is black.

Scientists have mimicked the structure of a chitin layer using ultramicroscopic structures—not nanotubes this time but cellulose nanofibrils, or microfibers. A layer of these fibrils is twenty or thirty times whiter than a sheet of paper. It's not the whitest white (some chemically based paints are whiter), but it's pretty close—and you can bet there will soon be other contenders.

History Mystery

*Did Benjamin Franklin fly a kite that was
struck by lightning?*

No, he did not.

Even though this is the one story that everyone knows about Benjamin
Franklin, there is still much mystery surrounding it. When did this experiment happen? How exactly was it set up? Did Franklin survive it? Is it even
possible to do what he claimed?

Here's the background. In the mid-1700s, Franklin was one of the world's
experts in electricity. He coined several words we still use today to describe
it, including "positive" and "negative," "charges," "conductor," and "battery."
He knew how to generate and store static electricity (the kind you produce
when you shuffle across a carpet in your socks). Franklin even hosted parties
where lines of people would join hands to let a shock run through them, to
the delight of onlookers.

Franklin suspected that lightning was similar to static electricity, in that a charge built up in a thundercloud and was released when the lightning completed a circuit from the cloud to the ground—or to a house, or sometimes, tragically, to a person. But how to prove it?

In most versions of the story, Franklin builds a simple kite with a pointed metal wire extending from it. (The wire was pointed to concentrate the electrical field around it.) The kite string was made of hemp with a metal key at the end. Below the key was a short piece of silk for Franklin to hold on to.

Franklin predicted that if he flew the kite into a cloud that had accumulated an electrical charge, that charge would jump to the point of the wire, run down the hemp string, and electrify the key. He assumed that the rain-soaked hemp would conduct electricity, but he planned to stand inside some sort of shelter to keep the short piece of silk dry. This would ensure that the charge wouldn't jump to Franklin himself, not so much to protect him but to ensure that the electricity would collect on the key.

Science _Fiction!_ Some people think that if they ride a motorcycle in a thunderstorm, the rubber tires will keep them safe. They won't! Rubber is indeed an insulator (meaning electricity passes through it only with difficulty), but that's not enough. In a car, you're protected not by the tires but by the metal body, which acts as what's called a Faraday cage. Any electricity from a lightning strike will pass around the metal and reach the ground, likely through the tires. There's no such protection on a motorcycle.

Is it safe to stand under a tree? No, it's not. In fact, it's less safe than being in the open. But don't lie prone on the ground either, thinking that a low profile will help. Any lightning that strikes the ground could easily pass through you, too. If you can't find shelter, squat down, tuck your head close to your knees, and wait out the storm.

As he waited, so the story goes, the strands of hemp suddenly stood up (the way your hair would if you touched a Van de Graaff generator at a science museum). Franklin tentatively extended a knuckle until it was close to but not touching the key, and a spark leaped across the gap. He'd demonstrated that the storm cloud had built up a static charge.

I feel a spark between us, don't you?

But this was a mild charge, a mere spark. If the clouds had actually generated a bolt of lightning, Franklin would almost certainly have been killed—which is exactly what happened when a Swedish physicist named Georg Richmann tried a similar experiment a few months later. So the straightforward answer to the original question is no, Franklin didn't fly a kite that was struck by lightning.

Science Fact! *Each year in North America, nearly five hundred people are struck by lightning, and on average, seventy of those are killed.*

But that's not the only question. When exactly did he do this? Franklin claimed he performed the kite experiment in June 1752. We know that a French team had done essentially the same experiment a month before, substituting a long iron rod for the kite. But Franklin wouldn't have heard of their efforts by June, given the slow pace of mail delivery in the eighteenth century, so he could at least claim he was as original as the Frenchmen. What's strange is that he didn't describe his experiment until October—long after he'd heard of what the French did. Was he copying them? Or did he actually perform the kite test in June?

Some people have argued that Franklin *never* flew the kite at all—that the whole thing was a complete hoax. In his book *Bolt of Fate*, Tom Tucker tries to reconstruct what Franklin is supposed to have done and uncovers several problems. For instance, Franklin said that to keep the strand of silk dry, the experimenter should be under some kind of roof. But Franklin also said that the kite had to be flown as high as possible to gather the electrical charge. It's nearly impossible to keep a kite string almost vertical and stand under a roof at the same time—unless there's a hole in the roof.

Tucker also argues that the kite described by Franklin was too small and fragile, and that the keys of the time were too heavy. And then there's the fact that Franklin waited months before revealing what he'd done, then described the feat in an oddly impersonal and vague way. All these things suggest—at least to Tucker—it just never happened.

Science _Fact!_ Yes, lightning can strike twice in the same place! Repeatedly, in fact. Tall buildings that reach closer to the clouds (like Franklin's kite) are especially susceptible to multiple strikes. The Empire State Building in New York was once hit eight times in less than half an hour.

Most Franklin experts, however, can't bring themselves to support the "hoax" theory. To them, it seems unlikely that the accomplished and respected Benjamin Franklin would stoop so low. And to what end? He'd admitted he was second to the French anyway, so he wasn't trying to claim he'd done something that had never before been done.

The doubts may never be completely erased, but the mysteries around Franklin's kite experiment haven't detracted from his deserved fame as a pioneer in the study of electricity.

People say I have
that shock factor.

Acknowledgments

This is the fifth in The Science of Why series, a collection of something like two hundred questions and answers. I've enjoyed writing them: even the simplest question seems inexorably to lead down a rabbit hole of more questions, more answers, more intrigue. It was a gift to be able to devote myself to this during the pandemic.

There are several people who have played a crucial role in the entire series: Kevin Hanson (whose goal of having it attract readers from ages "12 to infinity" seems to have been met), Nita Pronovost, Sarah St. Pierre, and Catherine Whiteside.

Janice Weaver edited this book and made that sometimes difficult process easy because her interest, even delight, in some of the stories shone through. Tony Hanyk's illustrations have uniquely added humor to the entire series, and Joanne O'Meara and Niki Wilson always stand by to correct, amend, research, and generally improve things.

My agent Jackie Kaiser and the people at Westwood Creative Artists have done a great job—as always—of ensuring that the books reach a substantial audience outside Canada.

And there are all the usual subjects who at least appear to be interested in this stuff, including the boys in the band, the men of the road, the Flatheaders, the Banff Centre winter group, and of course, the entire next generation of family.

And especially thanks to Mary Anne for providing warmth and support, enthusiasm, gentle criticism, and a science-y household.

Jay Ingram has written eighteen books, including the bestselling previous four books in this series, *The Science of Why*, *The Science of Why²*, *The Science of Why, Volume 3*, and *The Science of Why, Volume 4*. He was the host of Discovery Channel Canada's *Daily Planet* from the first episode until June 2011. Before joining Discovery, Ingram hosted CBC Radio's national science show, *Quirks & Quarks*. He has received the Sandford Fleming Medal & Citation from the Royal Canadian Institute, the Royal Society of Canada's McNeil Medal for the Public Awareness of Science, and the Michael Smith Award for Science Promotion from the Natural Sciences and Engineering Research Council of Canada. He is a distinguished alumnus of the University of Alberta, has received six honorary doctorates, and is a Member of the Order of Canada. Visit Jay at **JayIngram.ca.**

🐦 *@*jayingram

Got more questions that need answers?

Check out the first four volumes in this mega-selling series!

VOLUME 1

VOLUME 2

VOLUME 3

VOLUME 4

NATIONAL BESTSELLERS!